AF413265

Superinsulators, Bose Metals
and
High-T$_c$ Superconductors

The Quantum Physics of Emergent Magnetic Monopoles

Superinsulators, Bose Metals

and

High-T$_c$ Superconductors

The Quantum Physics of Emergent Magnetic Monopoles

Carlo A Trugenberger

SwissScientific Technologies SA, Switzerland

World Scientific

NEW JERSEY · LONDON · SINGAPORE · BEIJING · SHANGHAI · HONG KONG · TAIPEI · CHENNAI · TOKYO

Published by

World Scientific Publishing Co. Pte. Ltd.

5 Toh Tuck Link, Singapore 596224

USA office: 27 Warren Street, Suite 401-402, Hackensack, NJ 07601

UK office: 57 Shelton Street, Covent Garden, London WC2H 9HE

Library of Congress Control Number: 2022936285

British Library Cataloguing-in-Publication Data
A catalogue record for this book is available from the British Library.

ISBN 978-981-125-095-8 (hardcover)
ISBN 978-981-125-096-5 (ebook for institutions)
ISBN 978-981-125-097-2 (ebook for individuals)

For any available supplementary material, please visit
https://www.worldscientific.com/worldscibooks/10.1142/12688#t=suppl

Desk Editor: Ng Kah Fee

Typeset by Stallion Press
Email: enquiries@stallionpress.com

To my parents, who made it possible,
to my wife, who shared this journey,
to my kids, who embark on their own.

Preface

Physics courses go a long way to teach students that, while single electric charges are abundant, magnetic charges appear always in inseparable dipoles. Magnetic monopoles in a classical gauge theory would entail inadmissible singularities with infinite energy density. It took the genius of Dirac in the 1930s to show that quantum mechanics changes the game. He showed that, in a quantum theory of electromagnetism, these singularities are unobservable coordinate singularities if the product of all electric and magnetic charges is an integer multiple of the Planck constant h, the main numerical constant governing quantum behaviour. Quantum electromagnetism with magnetic monopoles is thus perfectly consistent.

Indeed, grand unified quantum field theories of the electromagnetic, weak and strong interactions do admit fundamental magnetic monopoles, in which the singularities are replaced by tiny inner cores containing very heavy elementary particles. Such fundamental monopoles are so heavy that they could have been produced only in the Big Bang. However, even after four decades of intensive and dedicated searches, no trace of them has ever been found.

This is not the end of the story though. Lighter Dirac monopoles could be realized as excitations in emergent condensed matter systems, where the typical energy scales are 12 to 13 orders of magnitude smaller than the grand-unified scale of all elementary interactions. Not only can they exist as isolated excitations, but, being light bosons, under appropriate conditions they can Bose condense. A condensate of electric charge is a superconductor, carrying current with no resistance. Correspondingly, a condensate of magnetic monopoles is the dual state, a *superinsulator*, opposing an infinite resistance to electric currents.

While a condensate shows the most dramatic consequences of emergent magnetic monopoles, single excitations also have an important effect. Indeed, they may constitute a trap for pairs of electrons, leading thus to a real-space-localized and very strong pairing mechanism which may be relevant for high-T_c superconductors.

Magnetic monopoles and related topological field theories have played a dominant role in my physics career, from my PhD thesis on monopole-induced proton decay catalysis to the first prediction of superinsulation in the mid-90s. In this book I have attempted to assemble a coherent picture of the physics of emergent Dirac monopoles in condensed matter systems, focusing on both the observed and the as yet potential new states of matter they induce.

A warm thank you goes to all collaborators who shared parts of this journey: I have learned a lot from each of them.

Carlo A. Trugenberger

Contents

Chapter 1

Introduction

Superinsulation is dual superconductivity. Already in the 70s, Nambu [1], Mandelstam [2] and 't Hooft [3] proposed their model of dual superconductivity as a possible explanation of quark confinement (for a review see [4]). The basic idea is that an Abelian magnetic monopole condensate in the $U(1)^{N-1}$ subgroup of SU(N) squeezes the lines of the colour-electric field into thin flux tubes, exactly as a charge condensate in a superconductor squeezes magnetic flux into vortices (for a review see [5]). This dual, colour-electric Meissner effect creates strings between the quarks at their endpoints. When quarks are pulled apart, it is energetically favourable to pull out of the vacuum additional quark-antiquark pairs and to form several short strings instead of a long string. As a consequence, colour charge can never be observed at distances above a fundamental length scale, $1/\Lambda_{QCD}$ and quarks are confined. Only colour-neutral hadron jets can be observed in collider events.

The idea that electric charge confinement can also be realized in emergent condensed matter systems, where it can lead to a new, *superinsulating* state of matter with infinite resistance even at finite temperatures was first proposed in [6] in the context of Josephson junction arrays (JJA) [7]. These are two-dimensional (2D) arrays of Josephson junctions and constitute a paradigmatic system for the observation of the superconductor-to-insulator (SIT) quantum transition [8,9]. At very low temperatures, when the junction islands turn superconducting, one could expect the whole array to show superconducting behaviour. However, depending on the ratio of the Josephson coupling and the charging energy of the junctions, the JJA can be in a global superconducting state or in an insulating state. In [6] it was shown that, near the quantum transition at zero temperature, this insulating state is actually a superinsulator with charge confinement. This idea was subsequently confirmed by a numerical approach [10].

JJA are metamaterials simulating the behaviour of thin superconducting films, in which the SIT can be driven by the thickness of the film, a magnetic field, gate voltages, or disorder. Twelve years after the first proposal, the idea of superinsulation was independently suggested as a new state of matter realized in thin superconducting films [11, 12] and experimentally detected in InO films [13] (albeit under a different name) and TiN films [11]. Finally, also the charge Berezinskii–Kosterlitz–Thouless (BKT) transition [14] into the infinite-resistance state was experimentally detected in NbTiN films [15].

While both early theoretical proposals [6] and [11, 12] considered logarithmic charge confinement, dual to logarithmic vortex confinement in the superconducting phase, it was recently realized [16] that the infinite-resistance state is caused by *linear confinement* of charges induced by magnetic monopoles [17], which are instantons in 2D [18, 20] and solitons in 3D [19, 20]. Contrary to charges, vortices are topological excitations, characterized by a topological quantum number. Therefore, contrary to charges, vortices are not conserved: they can appear and disappear from the vacuum in tunneling events that change the topological quantum number by one unit. Such tunneling events are called instantons [20] and, when they become frequent, they confine electric charges by 't Hooft's dual superconductivity mechanism [3]. In 3D, magnetic monopoles are solitons. Currents of such magnetic particles create circular electric fields and confine electric charges by the dual Maxwell–Ampère and Lorentz laws, which induce the dual Meissner effect. While magnetic monopoles appear in long-distance effective field theories of superconducting films [16], they can be derived directly from the microscopic model of JJA [21]. First hints of the electric Meissner effect have been detected in experiments on NbTiN films [22].

Charges in superinsulating films and JJA are thus linearly bound by an electric string, exactly as quarks within hadrons. Correspondingly, the observed quantum phase diagram around the SIT matches perfectly the quantum phase diagram of quantum chromodynamics (QCD), with a transition between confined hadronic matter and colour superconductivity [23]. Of course, the characteristic string scale is much larger than $1/\Lambda_{\rm QCD}$, the proton size, since the electromagnetic interaction is much weaker than the strong force. When samples are larger than this characteristic size, the string prevents charges to be separated and this results in the infinite resistance characterizing these materials. Small samples, instead, allow the direct investigation of asymptotic freedom effects [24] in the interior of an "electric pion" [22].

The SIT is one of the most representative examples of a quantum phase transition. Phase transitions are typically characterized by scaling and universality. The SIT is no exception, exactly at the transition point between superconductor and superinsulator it shows scaling and the system becomes a metal with one quantum of resistance [9], independently of material support. In [6] it was, however, predicted that, under certain conditions, this metallic point at the direct SIT could open up to a full fledged intermediate metallic phase. This metal made out of bosons was subsequently called a *Bose metal* [25–27].

The prediction of an intermediate anomalous metal phase was eventually verified experimentally, both in superconducting films [27] and in JJA [28–30]. However, the very concept of a metal in 2D is at clash with the orthodoxy that such a system is impossible because of Anderson localization [31] (for a review see [32]), let alone one whose constituents are bosons. The exact nature of the Bose metal remained correspondingly mysterious. There are, however, systems which prevent Anderson localization, even in low dimensions. These are topological insulators (for a review see [33]), in which charge transport is supported by egde modes protected from localization by symmetries induced by the connection with a topological bulk state. The Bose metal is a bosonic version of a topological insulator, a state of matter arising from the competition of two quantum BKT transitions for charges and vortices which are both kept out of condensate by strong quantum fluctuations [34]. In the direct SIT, instead, charges and vortices coalesce into an inhomogeneous texture of coexisting "bubbles" of charge and vortex condensates. This self-organized, emergent granularity is paradigmatic of the SIT [35].

The origin of the 2D SIT lies in the fact that the Coulomb interaction in a squeezed film becomes a screened logarithmic interaction. When the dielectric permittivity of the medium is large enough, the screening length can become larger than the sample size so that there is a competition between two purely logarithmic interactions: the electric Coulomb interaction and the magnetic logarithmic interaction between vortices. The three possible states of the system depend on which logarithmic interaction "wins out". There is, however, another possibility of creating a competition between electric and magnetic interactions. This arises in 3D when magnetic monopoles are present, this time the competition being between two $1/r$ potentials and leads to the 3D SIT.

Emergent magnetic monopoles (this time real "particles", not instantons) have been predicted to arise [36] in the recently observed 3D granular superconductors, made of localized "bubbles" of Bose condensate [37]. In this case, Josephson vortices (for a review see [5]) arise from non-trivial circulations of the order parameter on neighbouring bubbles. These Josephson vortices can be generically open, with magnetic monopoles and anti-monopoles at their end. When the tension of the vortices vanishes, small magnetic dipoles become free pairs of magnetic monopoles and anti-monopoles with a Dirac string between them. Since they are bosons, these monopoles are free to condense at low enough temperatures, leading to charge confinement, i.e. 3D superinsulation. Note that all the magnetic monopoles we will be discussing in this book are real, quantized Dirac monopoles, contrary to the monopole-like configurations at the end of very long aligned dipoles in spin ice [38] or the mirage monopoles hidden behind surfaces of topological insulators [39].

A particularly interesting class of 3D superinsulators are oblique superinsulators. They arise when a topological θ-term (for a review see [40]) is present in the electromagnetic response. Oblique superinsulators are condensates of *dyons*, magnetic monopoles that also carry electric charge, and confine only the excitations with the combination of electric and magnetic charge orthogonal to the condensate. Of outstanding relevance, in particular, are *strong superinsulators*, realized when $\theta = 2\pi$ and carrying a unit of both magnetic and electric charge. When the effective Coulomb interaction is large, strong superinsulators become topological states, supporting symmetry-protected fermionic surface states. Strong superinsulators are thus the 3D equivalent of Bose metals, in which a pure Fermi liquid behaviour of surface states is expected, protected by symmetry from impurity scattering and localization [41]. Strong superinsulators are thus the ideal candidate to describe the mysterious pseudogap state of high-T_c superconductors (for a review see [42]). They can also coexist with a Cooper pair condensate in a direct 3D SIT [41], giving rise to the superconducting dome of high-T_c materials, realized as a first-order quantum phase transition between dyon and pure charge condensates.

This picture of high-T_c superconductivity is predicated on the emergent granularity typical of the SIT. As a consequence it requires a real-space pairing mechanism to form bubbles of charge condensate. Emergent magnetic monopoles again come to the rescue and provide exactly such a mechanism [43]. Typical high-T_c superconductors are layered materials, in which a unit layer is made of two conducting CuO_2 planes, with possibly other atoms in between. A magnetic monopole of sufficiently high magnetic charge between the two conducting planes creates a potential well for electrons on the conducting planes near to it. Inter-plane magnetic monopoles are thus seeds for real-space Bose condensate bubbles of charge. These inter-layer magnetic monopoles can be effective, so-called pseudo-magnetic monopoles induced by curvature effects or hedgehog defects in magnetic order or real magnetic monopoles in loop-current ordered materials. Global superconductivity sets in when the density of monopoles is sufficiently high for paired charges to tunnel from one bubble to the other. We will provide an estimate showing that this new mechanism of superconductivity can survive at room temperature.

Magnetic monopoles, thus, while elusive as elementary particles in the universe, are realized as emergent condensed matter excitations, leading to new topological phases of matter and providing a possible explanation of the 40-year-old mystery of high-T_c superconductivity. The aim of this book is to review the key aspects and the present status of the quantum physics of these emergent magnetic monopoles, a fascinating field which is only in its infancy.

This book is written for graduate students and specialists interested in the application of field theory and topology methods to emergent condensed matter states. They are not designed to cover the entire field, which is a much too ambitious program, for which excellent books already exist. Rather, they are focused on the role of emergent magnetic monopoles, the new states of matter they induce and their possible star role in high-T_c superconductivity. The material presented in what follows presupposes a working knowledge of field theory and familiarity with basic concepts of condensed matter physics.

1.1. Units and notation

In the whole of this book we shall use natural units $\hbar = 1$, $c = 1$ and $\varepsilon_0 = 1$. We will restore physical units in some formulas, typically for comparison with experimental results. When we do so, we will always use SI units and we will explicitly make the reader attentive. The metric of space-time is chosen as $\mathrm{diag}(1, -1, -1, -1)$. Normal characters will be used to denote space-time variables, also Euclidean ones, boldface characters for purely spatial coordinates. Correspondingly, Greek letters are used for space-time indices, also Euclidean ones, and latin letters for pure space indices. Summation over Greek letters (Einstein notation) is always implied, unless otherwise specified. We shall indicate summation over latin spatial indices explicitly.

Chapter 2

Gauge theories and magnetic monopoles: A first encounter with superinsulators

Fundamental interactions are mediated by gauge fields. The key ingredient of the gauge principle is that a local action formulation of the interactions, leading to second-order equations of motion, requires the introduction of redundant, non-physical degrees of freedom which have to be accurately "removed" upon quantization. The theory is invariant under local transformations of the non-physical variables while the physical degrees of freedom are left untouched by these transformations. This is the meaning of local gauge invariance, which expresses a redundancy in the local description of physical phenomena rather than a true symmetry. The simplest example of a model with Abelian gauge symmetry is electromagnetism. The physical electric and magnetic fields are expressed via a four-vector gauge potential $A_\mu = (A_0 = V, -\mathbf{A})$ as,

$$\mathbf{E} = -\nabla V - \partial_0 \mathbf{A}, \quad \mathbf{B} = \nabla \times \mathbf{A}, \tag{2.1}$$

where V is the usual electric potential. They can be compacted into a two-tensor $F_{\mu\nu} = \partial_\mu A_\nu - \partial_\nu A_\mu$ so that $E^i = F^{i0}$ and $B^i = (1/2)\epsilon^{ijk}F_{jk}$. The Maxwell equations are expressed covariantly in terms of this tensor as

$$\partial_\mu F^{\mu\nu} = qej^\nu, \tag{2.2}$$

$$\partial_\mu \tilde{F}^{\mu\nu} = 0, \tag{2.3}$$

where e is the electron charge, q the integer charge unit of the theory, j^μ the number current and

$$\tilde{F}^{\mu\nu} = \frac{1}{2}\epsilon^{\mu\nu\alpha\beta}F_{\alpha\beta}, \tag{2.4}$$

is the dual electromagnetic tensor. The inhomogenous equations (2.2) are derived from the action

$$S = \int d^4x \left(\frac{-1}{4} F_{\mu\nu} F^{\mu\nu} - qeA_\mu j^\mu \right), \tag{2.5}$$

while the homogenous equations (2.3) are the Bianchi identities, following exclusively from the antisymmetry of the ϵ tensor.

Gauge invariance is the statement that both the electromagnetic field tensor $F^{\mu\nu}$ and the action (2.5) are left invariant by the transformations $A_\mu \to A_\mu + \partial_\mu \psi$. To extract the physical meaning of this, let us identify first the degrees of freedom of the model. We note that the component A_0 does not appear in the action with a time derivative. Therefore it is not a dynamical degree of freedom but, rather, only a Lagrange multiplier that enforces the Gauss law constraint $\partial_i E^i = qej^0$. The Gauss law (Poisson law for the electromagnetic potential) is, strictly speaking, not a dynamical equation of motion but, rather, a constraint on the configuration space, at classical level and on the Hilbert space, at quantum level. The dynamical degrees of freedom are the three spatial components A_i of the four-vector potential, which admit the Hodge decomposition

$$A_i = \partial_i \lambda + \epsilon^{ijk} \partial_j \tau_k, \tag{2.6}$$

into longitudinal and transverse components λ and τ_i, respectively. Gauge transformations affect only the longitudinal component λ. The electromagnetic fields that enter the action, however, do not depend on this longitudinal component. One might thus be tempted to eliminate it completely *ab initio* and this is indeed possible. The price to pay, however, is that the action becomes non-local. If we want to insist on a local action, which is crucial at quantum level, where not only the equations of motion matter, we are forced to introduce a redundant field λ. Since physics does not depend on this additional field we can transform it at will, even locally. Only the two transverse components of the vector potential $\mathbf{A}$ are dynamical degrees of freedom: these correspond to the two polarization states of a photon.

We will now show that what we just described is not entirely accurate. Let us consider the (λ, τ_i) decomposition of the magnetic field,

$$B^i = \epsilon^{ijk} \partial_j \partial_k \lambda + \left(-\nabla^2 \delta_{ij} + \partial_i \partial_j \right) \tau_j. \tag{2.7}$$

We have to distinguish carefully between a non-compact and a compact gauge symmetry [18]. In the former case, the gauge group is $\mathbb{R}$ and λ is an unbounded field. The first term in (2.7) then vanishes due to the antisymmetry of the ϵ tensor and, as expected, gauge invariance is respected. In the latter case, the gauge group is U(1), which is topologically equivalent to a circle S^1. The field λ is thus an *angular variable*, taking values in $[0, 2\pi r_G]$ where r_G is the radius of the gauge group. Enter the identity

$$\hat{\mathbf{z}}\, \delta^2(\mathbf{x}) = \frac{1}{2\pi}\epsilon^{3ij}\partial_i\partial_j\varphi, \tag{2.8}$$

where $\hat{\mathbf{z}}$ denotes a unit vector in the z direction, $\mathbf{x}$ are coordinates on the (x, y) plane, labeled by i and j and φ is the angular coordinate on this plane. As a consequence, the compact model admits topological vortex excitations in the form of closed or infinitely extended singular magnetic flux tubes carrying flux $2\pi r_G$. But these singularities are encoded in the longitudinal component λ which, according to the gauge principle, should never contribute to any physical observable. How can we save gauge invariance?

First of all we can exclude the support of these singularities from the domain of our model. This can be done in two ways [18]: either one considers the U(1) gauge group as arising by spontaneous symmetry breaking from a larger, non-Abelian compact group, such as O(3), or one formulates the theory on a lattice. In the former case the singularities are resolved by new degrees of freedom inside the symmetry breaking radius, in the latter case they reside in the centre of the lattice plaquettes and are thus not accessible by degrees of freedom living on the lattice. Either way, the compact model requires an ultraviolet (UV) cutoff to screen the singularities and is thus an effective field theory (when there is no dynamical matter coupled to it, as we will see below). Screening the singularities, however, is not enough at the quantum level, since magnetic flux tubes can be detected at long distances by the Aharonov–Bohm effect [44]. In quantum mechanics, when a particle of charge qe encircles a tube carrying flux Φ it acquires a phase factor $\exp(iqe\Phi)$. A charge cannot detect a flux tube by the Aharonov–Bohm effect if it satisfies the quantization condition $qe\Phi = 2\pi n$ with $n \in \mathbb{Z}$. Charges in a compact gauge theory are thus quantized. In the present case, requiring that the flux tubes singularities in our model cannot be detected even by the Aharonov–Bohm effect implies thus that q is quantized. Without restriction of generality we will choose q itself to be an integer, which identifies the coupling constant e as the electron charge. Then the gauge group radius is $r_G = 1/qe$. If these conditions are met, gauge invariance is guaranteed.

This is the end of the story if the vortices are closed or infinitely extended. But what about the case in which a vortex ends in a source point, which we will take to be the origin of our coordinate system for simplicity's sake? To satisfy the Maxwell equation $\nabla \cdot \mathbf{B} = 0$, the magnetic flux carried by the vortex has to flow out from the endpoint and it will do so in a radially symmetric way,

$$\mathbf{B} = \frac{(2\pi/qe)}{4\pi r^2}\hat{\mathbf{r}} + \frac{2\pi}{qe}\theta(-z)\delta^2(\mathbf{x})\,\hat{\mathbf{z}}, \tag{2.9}$$

where $\hat{\mathbf{r}}$ denotes a unit vector in the radial direction and we have assumed a straight vortex along the negative axis. However, since this vortex is unobservable by any physical process, this solution corresponds, for all purposes, to a fundamental *magnetic monopole* of strength

$$g = \Phi = \frac{2\pi}{qe}, \tag{2.10}$$

sitting at the origin, as shown in Fig. 2.1. Any other monopole must be a multiple of this unit. This is the famed Dirac quantization condition and the unobservable vortex attached to the monopole to ensure a divergenceless overall magnetic field is called the Dirac string, after P. M. A. Dirac, who first derived this solution in 1931 [45]. The vector potential corresponding

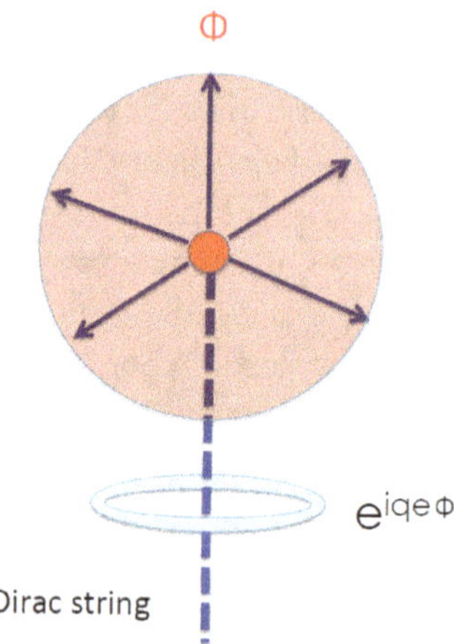

Fig. 2.1. A magnetic monopole of strength $g = \Phi$. A magnetic flux Φ is flowing up through the Dirac string and flows out radially from the monopole. If the Dirac quantization condition $qeg = 2\pi n$ is satisfied the Dirac string is unobservable even by quantum Aharonov–Bohm interferences.

to a unit magnetic monopole is

$$\mathbf{A} = \frac{1}{2qe} \frac{(1 - \cos\theta)}{\sin\theta} \nabla\varphi = \frac{g}{4\pi r} \frac{(1 - \cos\theta)}{\sin\theta} \hat{\varphi}, \qquad (2.11)$$

where $\hat{\varphi}$ is a unit vector in the φ direction [17].

Of course magnetic monopoles can also move, inducing thereby a magnetic current j_g^μ. When they are included in a compact U(1) gauge theory, the Maxwell equations must thus be altered to include the corresponding magnetic current,

$$\partial_\mu F^{\mu\nu} = qe j^\nu,$$
$$\partial_\mu \tilde{F}^{\mu\nu} = g j_g^\nu. \qquad (2.12)$$

We will see in a moment what are the consequences.

Magnetic monopoles have never been detected as elementary particles in isolation, even if electric charges are clearly quantized and quantum electrodynamics (QED) is thus the compact version. Indeed, magnetic monopoles do arise as solitons in grand unified theories (GUTs) such as SU(5) or O(10), in which the U(1) group appears as the lowest-energy surviving gauge interaction after spontaneous symmetry breaking [17]. Such monopoles have multiple structures. They are very heavy, with a mass of the order of the GUT gauge bosons (X) mass $m_\Phi \simeq m_X = O\left(10^{16}\right)$ GeV, this mass being mostly concentrated within the spontaneous symmetry breaking radius $R_X = 1/m_X$. Up to distances $O(1/m_Z)$, virtual W and Z bosons have sizeable effects and colour magnetic fields extend to even larger distances. Finally, at the longest distances only the U(1) Dirac magnetic charge survives. The heavy mass of monopoles explains why they cannot be produced in existing and planned accelerators. The only hope is to observe them in cosmic rays but, even in this case, no process occurring in the present universe would be sufficiently energetic to produce particles of such a huge mass. All monopoles present today, if ever, must have been produced in the early universe. However, although the standard cosmological scenario would allow an abundance of monopoles sufficient for their detection with current observational techniques, they have remained stubbornly elusive.

But what about magnetic monopoles in long distance *effective* QED theories of highly correlated condensed matter systems? In this case the UV cutoff would not be the X boson mass but a much lighter keV scale, corresponding to a nm size. Such magnetic monopoles would be lighter and fatter and could lead to observable effects in materials. Let us present a first, admittedly heuristic argument to show that magnetic monopoles can indeed have a large effect on emergent quantum states. To this end, let us assume an emergent ground state consisting of a condensate of magnetic monopoles and let us insert two external charges to probe the nature of this state. For simplicity of presentation, we will assume that the charges lie on the (x, y) plane and that there is a uniform distribution of steady orthogonal currents of j_g magnetic monopoles per unit time along the z direction. The dual Maxwell–Ampère equation (in the static limit $\partial_t B = 0$)

$$\nabla \times \mathbf{E} = -\frac{2\pi}{qe}\mathbf{j}_g \tag{2.13}$$

implies that these vertical monopole currents are surrounded by circular electric fields on the (x, y) plane, exactly as an electric current generates a circular magnetic field around itself. In the $A_0 = 0$ gauge, the solution of this equation for a distribution of monopole currents at points $\mathbf{x}_I$ on the (x, y) plane is given by

$$\begin{aligned} A_i &= \sum_I \frac{m_I}{qe} t j_g \, \partial_i \varphi_{\mathbf{x}_I}, \\ E^i &= \sum_I \frac{m_I}{qe} j_g \, \partial_i \varphi_{\mathbf{x}_I}, \end{aligned} \tag{2.14}$$

where, by subscript $\mathbf{x}_I$, we mean the phase coordinate based at this point and we have used the identity

$$\frac{1}{2\pi}\epsilon^{ij}\partial_i\partial_j\varphi_{\mathbf{x}_I} = \delta^2\left(\mathbf{x} - \mathbf{x}_I\right). \tag{2.15}$$

We have also assumed all currents to be equal, up to the directional sign $m_I = \pm 1$. The overall current in the sample is, of course, chosen to vanish, $\sum_I m_I = 0$.

The energy of this monopole current distribution is given by

$$H_{\mathrm{mon}} = \frac{1}{2}\int d^3\mathbf{x}\, \mathbf{E}^2(\mathbf{x}) = \frac{L_z}{2}\sum_{I,J}\frac{j_g m_I}{eq}\frac{j_g m_J}{eq}\int d^2\mathbf{x}\, \partial_i\varphi_{\mathbf{x}_I}\partial_i\varphi_{\mathbf{x}_J}, \tag{2.16}$$

where L_z denotes the sample extension in the z direction. We now use the identity (2.15) in the modified form

$$\partial_i\varphi_{\mathbf{x}_I} = 2\pi\epsilon^{ij}\frac{\partial_j}{-\nabla^2}\delta^2\left(\mathbf{x} - \mathbf{x}_I\right) \tag{2.17}$$

to rewrite the energy as

$$H_{\text{mon}} = \frac{L_z}{2} \sum_{I,J} \frac{j_g m_I}{eq} \frac{j_g m_J}{eq} \ln \frac{|\mathbf{x}_I - \mathbf{x}_J|}{\ell}, \qquad (2.18)$$

where we have used the 2D Poisson kernel

$$\frac{1}{-\nabla^2} \delta^2(\mathbf{x} - \mathbf{x}_I) = \frac{1}{2\pi} \ln \frac{|\mathbf{x} - \mathbf{x}_I|}{\ell}, \qquad (2.19)$$

normalized to vanish at the ultraviolet cutoff ℓ. The monopole currents are thus equivalent to a 2D neutral magnetic Coulomb gas [47] on the plane orthogonal to their direction.

Let us consider two widely separated probe charges on the x axis. In this dipole configuration there will be a high concentration of electric field lines on the x axis. Due to the dual Lorentz force

$$F = -\frac{2\pi m}{qe} \mathbf{v} \times \mathbf{E}, \qquad (2.20)$$

the monopoles, moving along the z direction, will experience a force along the y direction, in the positive or negative sense depending on the sign m of their charge. This means that the external probe charges will polarize the magnetic Coulomb gas so that all positive and negative charges lie in the same half-planes, above or below the x axis, depending on the sign of their charges. Recalling that these signs actually indicate the directions of the monopole currents along the z axis we obtain a configuration in which all circular electric fields due to these currents are in the clockwise sense in one half-plane and in the anticlockwise sense in the other half-plane. As a consequence, they will sum up along the x-axis segment between the two probe charges. The energy shift due to these external probe charges is thus

$$\Delta H = \sigma L, \qquad (2.21)$$

where L is the distance between them. This represents a *linear* potential with *string tension* σ. In other words, the monopole currents induce a squeezed electric flux tube, a string, between the two charges. Due to this string, a very large energy is required to separate the charge–anticharge pair. At large distances the interaction becomes so strong that it is energetically favourable to pop out of the vacuum a new charge–anticharge pair that forms two much shorter strings together with the original charges. This is the same mechanism that binds quarks together in hadrons [3,4] and that permits the observation of only hadron jets when the string is "pulled apart" in collider events. If a monopole condensate forms in a material, fundamental electric charges will be confined by the same mechanism and this confinement is the origin of a lack of current response to an applied voltage, i.e. of the infinite resistance of superinsulators.

Let us conclude this chapter by some more considerations about gauge theories that will come handy for what follows. First of all, in field theory, a factor e is often absorbed into the gauge potentials and fields, $eA_\mu \to A_\mu$, $eF_{\mu\nu} \to F_{\mu\nu}$, so that the action becomes

$$S = \int d^4x \left(\frac{-1}{4e^2} F_{\mu\nu} F^{\mu\nu} - qA_\mu j^\mu \right). \tag{2.22}$$

This has the advantage that, in the Euclidean formulation, obtained by the Wick rotation $x^0 \to -ix^0$, $A_0 \to iA_0$, the partition function

$$Z = \int \mathcal{D}A_\mu \; e^{-S_{\rm E}},$$

$$S_{\rm E} = \int d^4x \left(\frac{1}{4e^2} F_{\mu\nu} F_{\mu\nu} + iqA_\mu j_\mu \right), \tag{2.23}$$

takes the form of a statistical mechanics model in four dimensions, with the coupling constant e^2 playing the role of the effective temperature. Finally, we will mostly consider effective *non-relativistic* gauge theories in materials with a (relative) magnetic permeability $\mu = 1$ but a high (relative) dielectric permittivity ε, corresponding to a light velocity $v = 1/\sqrt{\varepsilon}$ in the medium. In this case the (Euclidean) gauge action in the original, non-rescaled fields becomes

$$S_{\rm E}^{\rm gauge} = \frac{1}{2} \int dvt \, d^3\mathbf{x} \left(\frac{1}{v^2} \mathbf{E}^2 - \mathbf{B}^2 \right), \tag{2.24}$$

which is the Wick rotated Legendre transform of the familiar expression

$$H_{\rm gauge} = \frac{v}{2} \int d^3\mathbf{x} \, (\mathbf{DE} + \mathbf{BH}), \tag{2.25}$$

for the energy of the electromagnetic field in a material, with $\mathbf{D} = \varepsilon\mathbf{E}$ and $\mathbf{B} = \mathbf{H}$ ($\mu = 1$). Note that, for dimensions bookkeeping simplicity, we choose the dimensions of the gauge fields to be $[A_0] = 1/\text{time}$, $[A_i] = 1/\text{space}$, in units $\hbar = 1$, $\varepsilon_0 = 1$, rather than the SI standard $[A_0^{\rm SI}] = 1/(\sqrt{v}\,\text{time})$, $[A_i^{\rm SI}] = 1/(\sqrt{v}\,\text{space})$. By absorbing again a factor e into the gauge fields we get

$$S_E^{nr} = \int dvt \, d^3x \left[\frac{1}{2e^2} \left(\frac{1}{v^2} \mathbf{E}^2 - \mathbf{B}^2 \right) + i\frac{q}{v} A_\mu j_\mu \right]. \tag{2.26}$$

We can now formally define a new variable $x^0 = vt$ (actually it is a space variable), a corresponding derivative $\partial_0 = (1/v)\, \partial/\partial t$ and current $\mathbf{j} \to (1/v)\mathbf{j}$ and a rescaled non-dynamical field $A_0 \to (1/v)A_0$. In these new variables, the action takes the form

$$S_E^{nr} = \int d^4x \left(\frac{1}{4e^2} F_{\mu\nu} F^{\mu\nu} + iqA_\mu j_\mu \right),\tag{2.27}$$

which formally coincides with the $c = 1$ relativistic case (2.23) formulated on an anisotropic four-space with one of the coordinates rescaled by a numerical factor $v < 1$. It is this version of the model that we will mostly use in what follows.

Chapter 3

Gauge theories in 2+1 dimensions: The Chern–Simons term

In (2+1) dimensions the Maxwell action is power-counting irrelevant, since the coupling constant e^2 has dimension [mass]. Correspondingly, the model is super-renormalizable and can be expected to be plagued by severe infrared divergences if regulating mass terms are absent. Mass terms, however, typically break gauge invariance. In a series of seminal papers [48], Deser, Jackiw and Templeton showed, on the contrary, that a gauge invariant mass term exists in (2+1) dimensions, if one is willing to pay the price of violating the discrete symmetries of parity $\mathcal{P}$, with group $\mathbb{Z}_2^P$, and time-reversal $\mathcal{T}$, with group $\mathbb{Z}_2^T$. This is the famous Chern–Simons (CS) term and the augmented massive electrodynamics has the Lagrangian density

$$\mathcal{L} = \mathcal{L}_{\mathrm{M}} + \mathcal{L}_{\mathrm{CS}} + \mathcal{L}_{\mathrm{coupling}} = \frac{-1}{4e^2} F_{\mu\nu} F^{\mu\nu} - \frac{k}{4\pi} A_\mu \epsilon^{\mu\alpha\nu} \partial_\alpha A_\nu - q A_\mu j^\mu.$$

$$(3.1)$$

Let us focus first on the pure gauge action, leaving out, for the moment, the coupling to charges. Under a gauge transformation $A_\mu \to A_\mu + \partial_\mu \lambda$ the Lagrangian density (3.1) changes by a total derivative,

$$\mathcal{L} \to \mathcal{L} - \partial_\mu \left(\frac{k}{4\pi} \epsilon^{\mu\alpha\nu} \partial_\alpha A_\nu \right), \qquad (3.2)$$

so that the corresponding action is gauge invariant on infinite spaces, on which the electromagnetic fields $\mathbf{E}$ and B (the magnetic field is a pseudoscalar in two dimensions) vanish at infinity for finite energy solutions. On finite bounded spaces and compact spaces more care is needed, we will treat these cases below. Poincaré invariance is obvious. Actually, the Chern–Simons action enjoys a much larger invariance under general coordinate transformations. This is realized in a trivial way, namely by metric independence of the action. Suppose we want to define the Chern–Simons action on a curved manifold with metric $g_{\mu\nu}$: in this case we have to substitute the antisymmetric symbol $\epsilon^{\mu\alpha\nu}$ with the corresponding contravariant tensor $\epsilon^{\mu\alpha\nu}/\sqrt{g}$, where $g = -\det g_{\mu\nu}$ [49]. This ensures that the contraction with the covariant vectors A_μ, ∂_α and A_ν produces a scalar. Moreover, we must also substitute d^3x with the appropriate volume form $\sqrt{g}\, d^3x$, thereby obtaining

$$S_{\text{CS}} = \int \sqrt{g}d^3x \ A_\mu \frac{1}{\sqrt{g}}\epsilon^{\mu\alpha\nu}\partial_\alpha A_\nu. \tag{3.3}$$

It is evident from this expression that all metric dependence cancels out. Therefore, the Chern–Simons term does not depend on the geometry of the manifold on which it is defined but only on its topology: this is why it is called a *topological term*. A further consequence of the metric independence is that the Chern–Simons energy-momentum tensor, defined as the functional derivative of the action (3.3) with respect to the metric $g_{\mu\nu}$, vanishes identically. Therefore the Chern–Simons contribution to the Hamiltonian also vanishes and this retains its positive definite Maxwell form (2.25), ensuring that the excitations are not tachyonic.

The Chern–Simons term is invariant under the discrete charge conjugation transformation $\mathcal{C}$, $A_\mu \to -A_\mu$, but changes sign under the (2+1)-dimensional parity transformation $\mathcal{P}$,

$$\begin{aligned}
\mathbf{x} &= \left(x^1, x^2\right) \to \mathbf{x}' = \left(-x^1, x^2\right), \\
A_0\left(t, \mathbf{x}\right) &\to A_0\left(t, \mathbf{x}'\right), \\
A_1\left(t, \mathbf{x}\right) &\to -A_1\left(t, \mathbf{x}'\right), \\
A_2\left(t, \mathbf{x}\right) &\to A_2\left(t, \mathbf{x}'\right),
\end{aligned} \tag{3.4}$$

and under the time-reversal transformation $\mathcal{T}$,

$$\begin{aligned}
A_0\left(t, \mathbf{x}\right) &\to A_0\left(-t, \mathbf{x}\right), \\
A_1\left(t, \mathbf{x}\right) &\to -A_1\left(-t, \mathbf{x}\right), \\
A_2\left(t, \mathbf{x}\right) &\to -A_2\left(-t, \mathbf{x}\right).
\end{aligned} \tag{3.5}$$

Therefore, the Chern–Simons coupling constant is $\mathcal{P}$ and $\mathcal{T}$ odd; however, the combined $\mathcal{PT}$ transformation leaves the Chern–Simons term invariant, so that $\mathcal{CPT}$ symmetry is preserved.

A notable property of the Chern–Simons term is that it contains one less derivative than the standard Maxwell term, which makes it power-counting marginal, with a dimensionless coupling constant k. Therefore it will dominate in the infrared (IR) region and determine the large-distance properties of the field theory. We can thus expect modifications to the standard properties of electromagnetism in the low-energy behaviour, exactly where the potential infrared divergences would cause problems. To show that this is indeed so let us derive the pure gauge field equations corresponding to the Lagrangian density (3.1),

$$\partial_\nu F^{\mu\nu} - \left(\frac{ke^2}{2\pi}\right)\epsilon^{\mu\alpha\beta}\partial_\alpha A_\beta = 0. \tag{3.6}$$

These can be read also as field equations for the dual field strength

$$F^\mu = \frac{1}{2}\epsilon^{\mu\alpha\beta}F_{\alpha\beta}, \tag{3.7}$$

which, in (2+1) dimensions, is a three-pseudovector,

$$\epsilon^{\mu\nu}_\alpha \partial_\nu F^\alpha + \left(\frac{ke^2}{2\pi}\right)F^\mu = 0. \tag{3.8}$$

By contracting with $\epsilon_{\mu\gamma\delta}\partial^\delta$ one then obtains the second-order field equation

$$\left(\Box + m^2\right)F^\mu = 0,$$
$$m = \frac{|k|e^2}{2\pi}, \tag{3.9}$$

which clearly shows the *massive* character of the one-particle states. These are called *topologically massive photons* due to the topological nature of the mass term in the Lagrangian. They provide a good example of the general rule that gauge invariance alone is not sufficient to prevent a photon mass. In condensed matter applications, this means that superconductivity, in the sense of dissipationless transport and the Meissner effect, is not tied to spontaneous symmetry breaking and the Anderson–Higgs mechanism [50, 51]. We will see below that this is indeed realized in two-dimensional (2D) superconductivity.

The spin of the topologically massive photons can be inferred from the representation of the (2+1)-dimensional Poincaré algebra

$$[J^\mu, J^\nu] = i\epsilon^{\mu\nu\alpha} J_\alpha,$$

$$[J^\mu, P^\nu] = i\epsilon^{\mu\nu\alpha} P_\alpha, \tag{3.10}$$

$$[P^\mu, P^\nu] = 0,$$

expressed in terms of

$$J^\mu = \frac{1}{2}\epsilon^{\mu\alpha\beta} L_{\alpha\beta} = \left(-R, B^2, -B^1\right), \tag{3.11}$$

where $L_{\mu\nu}$ generate infinitesimal Lorentz transformations SO(2,1), with standard basis given by the (unique) rotation R and the two boosts B^i, and P_μ generate translations. Relativistic one-particle states must carry a unitary, irreducible representation of the universal covering of the Poincaré group [52]. Such representations are classified by the eigenvalues of two Casimir invariants. In (2+1) dimensions these are P^2 and the Pauli–Lubanski pseudoscalar $P \cdot J = P_\mu \epsilon^{\mu\alpha\beta} L_{\alpha\beta}$. Their eigenvalues m^2 and $(-sm)$ define the mass m and the spin s of the representation. The adjoint representation of the SO(2,1) Lorentz algebra is

$$\left(J^\alpha\right)_{\mu\nu} = i\epsilon_\mu{}^\alpha{}_\nu. \tag{3.12}$$

Let us now Fourier transform the field equation (3.8),

$$-i\epsilon^{\mu\nu}_\alpha \, p_\nu F^\alpha(p) + \left(\frac{ke^2}{4\pi}\right) F^\mu(p) = 0. \tag{3.13}$$

Using (3.9) and (3.12) this can be rewritten as

$$(p \cdot J)^\mu_\alpha F^\alpha(p) = -\frac{k}{|k|} m F^\mu(p), \tag{3.14}$$

which we recognize as the Pauli–Lubanski condition for a vector representation of mass m and spin ± 1. Topologically massive photons are thus vector particles with a spin given by the sign of the Chern–Simons term,

$$s = \frac{k}{|k|}. \tag{3.15}$$

Can one formulate a $\mathcal{P}$- and $\mathcal{T}$-preserving version of the topological Chern–Simons term? The answer is affirmative, provided one introduces two gauge fields A_μ and B_μ with the Lagrangian density

$$\mathcal{L} = \frac{-1}{4e^2} F_{\mu\nu} F^{\mu\nu} - \frac{k}{2\pi} A_\mu \epsilon^{\mu\alpha\nu} \partial_\alpha B_\nu + \frac{-1}{4g^2} G_{\mu\nu} G^{\mu\nu}, \qquad (3.16)$$

where $G_{\mu\nu} = \partial_\mu B_\nu - \partial_\nu B_\mu$ is the field strength associated to the second gauge field B_μ and g^2 is an additional coupling with dimension [mass]. The second term in (3.16) defines a *mixed,* or *doubled* Chern–Simons term. If B_μ is taken as a pseudovector field, transforming under $\mathcal{P}$ and $\mathcal{T}$ as in (3.4) and (3.5) but with an overall minus sign, then the doubled model (3.16) preserves these discrete symmetries. Obviously, there are now two independent Abelian gauge symmetries. Repeating the derivation leading to (3.9) one obtains the equations of motion

$$\left(\Box + m^2 \right) F^\mu = 0,$$
$$\left(\Box + m^2 \right) G^\mu = 0, \qquad (3.17)$$
$$m = \frac{|k| eg}{2\pi},$$

where $G^\mu = \epsilon^{\mu\alpha\beta} \partial_\alpha B_\beta$ is the dual field strength of the second gauge field. Both gauge fields acquire thus the same topological mass which, in this case, is determined by the product of the two couplings.

Let us now couple the two gauge fields to point sources,

$$\mathcal{L} \to \mathcal{L} - A_\mu j^\mu - B_\mu \phi^\mu, \qquad (3.18)$$

where

$$j^\mu = \sum_I q_I \int ds \, \frac{dx_I^\mu}{ds} \, \delta^3 \left(x - x_I(s) \right),$$
$$\phi^\mu = \sum_J \Phi_J \int dt \, \frac{dx_J^\mu}{dt} \, \delta^3 \left(x - x_J(t) \right), \qquad (3.19)$$

are assumed conserved, $\partial_\mu j^\mu = 0$, $\partial_\mu \phi^\mu = 0$, and $x_I(s)$ and $x_J(t)$ parametrize the closed trajectories. We would like to obtain the effective action for the sources, induced by integrating over the gauge fields mediating the interactions. To do so we will introduce a technique which we will use repeatedly over the course of this book, namely the Wick rotation to Euclidean space, which we have already briefly mentioned at the end of Chapter 2. This amounts to introducing an imaginary time variable $t \to -it$ and a corresponding transformation $A_0 \to iA_0$ so that we can transform the model into a statistical mechanics model defined on a

Euclidean 3-space with metric $\mathrm{diag}(1,1,1)$, on which we do not have to distinguish anymore between upper and lower indices of vectors. The effective action is defined then by

$$e^{-S_{\mathrm{eff}}(q_I,\Phi_J)} = \frac{1}{Z\left(A_\mu, B_\mu\right)} \int \mathcal{D}A_\mu \mathcal{D}B_\mu e^{-S_{\mathrm{E}}(A_\mu,B_\mu)+iA_\mu j_\mu + iB_\mu \phi_\mu},$$

$$S_{\mathrm{E}}\left(A_\mu, B_\mu\right) = \int d^3x \left(\frac{1}{4e^2}F_{\mu\nu}F_{\mu\nu} + i\frac{k}{2\pi}A_\mu \epsilon_{\mu\alpha\nu}\partial_\alpha B_\nu + \frac{1}{4g^2}G_{\mu\nu}G_{\mu\nu}\right),$$

$$\tag{3.20}$$

where $Z\left(A_\mu, B_\mu\right)$ is the partition function of the mixed Chern–Simons model.

In order to perform the Gaussian integration required in this computation, some care is needed, since the Maxwell kernel

$$M_{\mu\nu} = -\nabla^2 \delta_{\mu\nu} + \partial_\mu \partial_\nu \tag{3.21}$$

has a zero mode due to gauge invariance and cannot be inverted. As usual in gauge theories, one introduces a gauge fixing term to eliminate this zero mode,

$$M_{\mu\nu} \to M_{\mu\nu}^{\xi} = -\nabla^2 \delta_{\mu\nu} + (1+\xi)\partial_\mu \partial_\nu. \tag{3.22}$$

This modified kernel can now be inverted,

$$\left(M_{\mu\nu}^{\xi}\right)^{-1} = \frac{1}{-\nabla^2}\left(\delta_{\mu\nu} + \frac{1+\xi}{\xi}\frac{\partial_\mu \partial_\nu}{-\nabla^2}\right). \tag{3.23}$$

As expected, the inverted modified kernel blows up for $\xi \to 0$. However, the diverging term vanishes because of current conservation. One can thus safely take the limit $\xi \to 0$ when the inverted kernel is applied on conserved currents. Using this procedure we obtain the effective action

$$S_{\mathrm{eff}} = \int d^3x \left(\frac{e^2 k^2}{2} j_\mu \frac{\delta_{\mu\nu}}{m^2 - \nabla^2} j_\nu + \frac{g^2 k^2}{2}\phi_\mu \frac{\delta_{\mu\nu}}{m^2 - \nabla^2}\phi_\nu \right.$$

$$\left. + i\frac{2\pi}{k}m^2 j_\mu \frac{\epsilon^{\mu\alpha\nu}\partial_\alpha}{\nabla^2(m^2 - \nabla^2)}\phi_\nu \right). \tag{3.24}$$

This expression becomes particularly interesting in the $m \to \infty$ limit, in which the gauge action becomes a topological mixed Chern–Simons term only,

$$S_{\mathrm{eff}}\left(q_I, \Phi_J\right) = i\frac{2\pi}{k}\int d^3x\; j_\mu \epsilon^{\mu\alpha\nu}\frac{\partial_\alpha}{\nabla^2}\Phi_\nu. \tag{3.25}$$

Using the 3D Green function

$$\frac{1}{-\nabla^2}\delta^3(x) = \frac{1}{4\pi}\frac{1}{|x|}, \tag{3.26}$$

we obtain

$$S_{\text{eff}}(q_I, \Phi_J) = i\frac{2\pi}{k}\sum_{IJ} q_I \Phi_J \Phi(C_I, C_J),$$

$$\Phi(C_I, C_J) = \frac{1}{4\pi}\int_0^1 ds \int_0^1 dt\, \frac{dx_I^\mu}{ds}\epsilon^{\mu\alpha\nu}\frac{(x_I - x_J)^\alpha}{|x_I - x_J|^3}\frac{dx_J^\nu}{dt}. \tag{3.27}$$

The quantity $\Phi(C_I, C_J)$ is the Gauss linking number of the two closed curves C_I and C_J, the simplest topological knot invariant (for a review see [53]), see Fig. 3.1. The phase factor $\exp iS_{\text{eff}}(q_I, \Phi_J)$ is the description of the topological Aharonov–Bohm [44] and Aharonov–Casher [54] phases in the Euclidean path integral formulation of the quantum mechanics of point charges and vortices. As was pointed out by Wilczek [55], these topological phases have a local formulation by coupling the charge and vortex trajectories to two gauge fields with a mixed Chern–Simons interaction. In this representation, the topological mixed Chern–Simons term is the infrared-dominant marginal term, while the two Maxwell terms are super-renormalizable, having coupling constants with dimension [mass]. In the

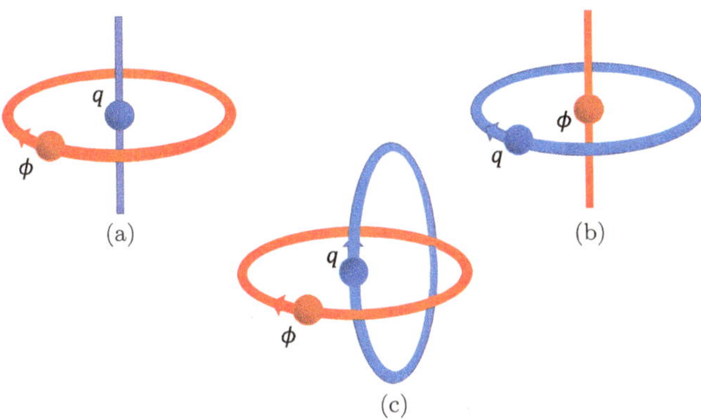

Fig. 3.1. Schematic examples of Aharonov–Bohm and Aharonov–Casher linkings of charges and vortices: (a) Aharonov–Casher effect of a vortex encircling a static charge, (b) Aharonov–Bohm effect of a charge encircling a static vortex, (c) combined effect. From M. C. Diamantini, A. Yu. Mironov, S. V. Postolova, X. Liu, Z. Hao, D. M. Silevitch, Ya. Kopelevich, P. Kim, C. A. Trugenberger and V. M. Vinokur; Bosonic topological intermediate state in the superconductor–insulator transition. *Phys. Lett.* **A384**, 126570 (2020), ©Elsevier (2020).

long-distance limit $m \to \infty$, it appears that only a purely topological theory survives. As we will show below, however, the limit $m \to \infty$ does not commute with quantization since it involves a phase-space reduction [56]. For physical applications (as opposed to purely mathematical ones concerning knot theory), the topological theory *always* has to be considered as the $m \to \infty$ limit of a topologically massive one since, otherwise, physical states are not normalizable. When the gauge symmetries are compact, a non-trivial dependence of the ground state on the dimensionless ratio g/e survives the limit $e \to \infty$ and $g \to \infty$.

To conclude this chapter we would like to retain the (Euclidean) non-relativistic form of the mixed topologically massive gauge theory, corresponding to (2.26),

$$S_{\mathrm{E}}^{\mathrm{nr}} = \int dt d^2 x \left(i\frac{k}{2\pi} A_\mu \epsilon_{\mu\alpha\nu} \partial_\alpha B_\nu + \frac{v}{2e^2} F_0 F_0 + \frac{1}{2e^2 v} F_i F_i \right.$$

$$\left. + \frac{v}{2g^2} G_0 G_0 + \frac{1}{2g^2 v} G_i G_i + iq A_\mu j_\mu + i\Phi B_\mu \phi_\mu \right), \qquad (3.28)$$

where $v = 1/\sqrt{\varepsilon}$ is the velocity of light in the medium and the coupling constants e^2 and g^2 have now dimension [1/length]. After the same rescaling leading to (2.27) we get back

$$S_{\mathrm{E}}^{\mathrm{nr}} = \int d^3 x \left(i\frac{k}{2\pi} A_\mu \epsilon_{\mu\alpha\nu} \partial_\alpha B_\nu + \frac{1}{2e^2} F_\mu F_\mu \right.$$

$$\left. + \frac{1}{2g^2} G_\mu G_\mu + iq A_\mu j_\mu + i\Phi B_\mu \phi_\mu \right), \qquad (3.29)$$

where, again, all gauge field and derivative components with an index "0" and all spatial currents are rescaled by a factor $(1/v)$.

Chapter 4

Lattice Chern–Simons term

In the following chapters we shall consider compact Abelian theories, with gauge group U(1) rather than $\mathbb{R}$. As was pointed out by Polyakov [19], a proper formulation of a compact U(1) gauge theory requires either spontaneous symmetry breaking of a compact non-Abelian group or the introduction of an ultraviolet lattice regularization. We shall choose the latter procedure. Define ℓ as the spacing on a hypercubic lattice with sites denoted by $\{x\}$ and directions indicated by Greek letters and introduce the forward and backward derivatives and shift operators

$$
\begin{aligned}
d_\mu f(x) &= \frac{f(x + \ell\hat{\mu}) - f(x)}{\ell}, \quad S_\mu f(x) = f(x + \ell\hat{\mu}), \\
\hat{d}_\mu f(x) &= \frac{f(x) - f(x - \ell\hat{\mu})}{\ell}, \quad \hat{S}_\mu f(x) = f(x - \ell\hat{\mu}).
\end{aligned}
\tag{4.1}
$$

Summation by parts on the lattice interchanges both the two derivatives (with a minus sign) and the two shift operators. Gauge transformations are defined by using the forward lattice derivative. In the case of non-relativistic theories, the lattice is anisotropic and we shall denote by ℓ_0 the lattice spacing in the time direction. For later convenience we shall also define the finite difference operators

$$
\begin{aligned}
\Delta_\mu f(x) &= f(x + \ell\hat{\mu}) - f(x), \\
\hat{\Delta}_\mu f(x) &= f(x) - f(x - \ell\hat{\mu}).
\end{aligned}
\tag{4.2}
$$

These are also interchanged when summing by parts on the lattice.

We will also need to formulate mixed Chern–Simons gauge theories on a 3D Euclidean lattice. In doing so, some care is needed when defining the lattice analogue of the Chern–Simons operator $\epsilon_{\mu\alpha\nu}\partial_\alpha$. Indeed, the naïve substitution of the continuum derivative with a lattice derivative leads to a mixed Chern–Simons term which is not gauge invariant. Following [6] we define, instead, the two lattice Chern–Simons operators

$$k_{\mu\nu} = S_\mu \epsilon_{\mu\alpha\nu} d_\alpha, \quad \hat{k}_{\mu\nu} = \epsilon_{\mu\alpha\nu} \hat{d}_\alpha \hat{S}_\nu, \tag{4.3}$$

where no summation is implied over equal indices. Summation by parts on the lattice interchanges also these two operators, but without a minus sign. Gauge invariance of the mixed Chern–Simons term $\sum_x A_\mu k_{\mu\nu} B_\nu$ is then guaranteed by the relations

$$k_{\mu\nu} d_\nu = \hat{d}_\mu k_{\mu\nu} = 0, \quad \hat{k}_{\mu\nu} d_\nu = \hat{d}_\mu \hat{k}_{\mu\nu} = 0. \tag{4.4}$$

The product of the two Chern–Simons terms gives the lattice Maxwell operator

$$k_{\mu\alpha} \hat{k}_{\alpha\nu} = \hat{k}_{\mu\alpha} k_{\alpha\nu} = -\delta_{\mu\nu} \nabla^2 + d_\mu \hat{d}_\nu, \tag{4.5}$$

where $\nabla^2 = \hat{d}_\mu d_\mu$ is the 3D Laplacian. A discrete dual field strength is also formulated via the operator $k_{\mu\nu}$,

$$F_\mu = k_{\mu\nu} A_\nu. \tag{4.6}$$

Finally, we also define the finite difference Chern–Simons operators

$$K_{\mu\nu} = S_\mu \epsilon_{\mu\alpha\nu} \Delta_\alpha, \quad \hat{K}_{\mu\nu} = \epsilon_{\mu\alpha\nu} \hat{\Delta}_\alpha \hat{S}_\nu. \tag{4.7}$$

For simplicity of notation we will use the same symbol ∇^2 also for the finite difference Laplacian. It will always be clear from context if we mean the discrete Laplacian or the dimensionless finite difference Laplacian.

Chapter 5

Saddle points, topological excitations and instantons

Statistical mechanics and quantum mechanics both deal with fluctuations, thermal and quantum respectively. Formally, their partition functions,

$$Z_{\text{stat.}} = \int_{-\pi}^{+\pi} \mathcal{D}\varphi \, e^{-\beta H(\varphi)}, \tag{5.1}$$

$$Z_{\text{quant.}} = \int_{-\pi}^{+\pi} \mathcal{D}\varphi \, e^{-g S_{\text{E}}(\varphi)}, \tag{5.2}$$

with H the Hamiltonian and S_{E} the Euclidean action, both formulated in terms of a set of angular fields φ, look the same. Here $\beta = 1/T$ is the inverse temperature (we set the Boltzmann constant $k_{\text{B}} = 1$) and g is the quantum coupling constant. If we extract the typical energy scale J in the statistical mechanics problem, then g plays exactly the same role as J/T in statistical mechanics. The only difference is that the statistical mechanics problem is defined in a D-dimensional space, whereas the quantum mechanical problem is formulated on a $(D+1)$-dimensional Euclidean space-time.

In this chapter we shall be interested in situations in which the above functional integrals are dominated by saddle points. In particular, we shall be interested in localized saddle point solutions that cannot be smoothly connected to the trivial minimum $\varphi = 0$. Such saddle points are characterized by a topological quantum number. In the case of statistical mechanics, they constitute the *topological excitations* of the model and can be point-like, string-like or even localized on higher-dimensional hypersurfaces, in which case they go under the name of p-branes. In case of the quantum mechanics problem we shall focus mostly on point-like saddle points, although we will give also an example of an extended solution. Point-like saddle points of the action with a unit topological quantum number interpolate, in Euclidean time, between two topological states differing exactly by this unit of the same topological quantum number. They represent thus quantum tunneling events between two ground states, called topological vacua (or also superselection sectors) differing by one unit of the topological quantum number. Such point-like saddle points of the action are called *instantons* (for a review on topological excitations and instantons see [20]).

5.1.　The XY model and vortices

Consider a model of phases $\varphi_{\mathbf{x}} \in [0, 2\pi]$ defined on the sites $\mathbf{x}$ of a 2D lattice and governed by the Hamiltonian

$$H = J \sum_{\mathbf{x},i} \left(1 - \cos\left(\Delta_i \varphi_{\mathbf{x}}\right)\right). \tag{5.3}$$

This planar rotor model has a global O(2) symmetry under simultaneous rotation of all phases by the same angle. It is conventionally called the XY model because it describes rotors in the (x, y) plane. The ground state configurations of the phases are such that $\cos\left(\Delta_i \varphi_{\mathbf{x}}\right) = 1$, $\forall \mathbf{x}, i$. For low temperatures, phase configurations will be peaked about these ground states and we can substitute the partition function with

$$Z = \sum_{\{n_{\mathbf{x},i}\}} \int_{-\pi}^{+\pi} \mathcal{D}\varphi \, \mathrm{e}^{-\frac{\beta J}{2} \sum_{\mathbf{x},i} \left(\Delta_i \varphi_{\mathbf{x}} - 2\pi n_{\mathbf{x},i}\right)^2} \tag{5.4}$$

where $\{n_{\mathbf{x},i}\}$ are a set of integers defined on the links i at site $\mathbf{x}$ of the lattice and $\mathcal{D}\varphi = \prod_{\mathbf{x}} d\varphi_{\mathbf{x}}$. This procedure, valid at low enough temperatures, is called the Villain approximation (for a review see [57]).

The important point is that, due to the periodicity of the phases, the ground state condition $\cos\left(\Delta_i \varphi_{\mathbf{x}}\right) = 1$ fixes the phases only up to 2π rotations. The contour of a lattice plaquette is topologically equivalent to a circle S^1. Since the first homotopy group of the symmetry group is $\Pi_1\left(O(2)\right) = \mathbb{Z}$, each plaquette P is characterized by an integer topological quantum number N_P describing how many times the phases wind around the contour of the plaquette. When this winding $N_P = \pm 1$ we have a fundamental *vortex* on the plaquette, a point-like (in 2D) topological excitation that cannot be removed by small deformations of the phases. The integer $N = \sum_P N_P$, representing the total number of vortices, is the topological quantum number of the ground state. Ground states with different N cannot be connected by small deformations of the phases and are the different superselection sectors, or topological vacua, of the model.

Following [19], we decompose the integers $\{n_{\mathbf{x},i}\}$ in their transverse and longitudinal components according to

$$n_{\mathbf{x},i} = \Delta_i k_{\mathbf{x}} + \Delta_i \xi_{\mathbf{x}} + \epsilon_{ij} \Delta_j \lambda_{\mathbf{x}}, \tag{5.5}$$

where $\{k_{\mathbf{x}}\}$ are integers and $\nabla^2 \lambda_{\mathbf{x}} = m_{\mathbf{x}} \in \mathbb{Z}$ represent the integer vortex degrees of freedom on the sites of the lattice. We have thus replaced two integers $\{n_{\mathbf{x},i}\}$, $i = 1,2$ with two integers $\{k_{\mathbf{x}}\}$ and $\{m_{\mathbf{x}}\}$. The integers $\{k_{\mathbf{x}}\}$ can now be used to shift the integration domain of the phases from $[-\pi, +\pi]$ to $[-\infty, +\infty]$ and the real variables $\{\xi_{\mathbf{x}}\}$ can then also be reabsorbed into the resulting real phases. The partition function decomposes thus into the product of Gaussian fluctuations and a topological excitations contribution,

$$Z = Z_0 \cdot Z_{\text{top}} = \int_{-\infty}^{+\infty} \mathcal{D}\varphi \, e^{-\frac{\beta J}{2} \sum_{\mathbf{x},i} (\Delta_i \varphi_{\mathbf{x}})^2}$$

$$\cdot \sum_{\{m_{\mathbf{x}}\}} e^{-2\pi^2 J\beta \sum_{\mathbf{x}\mathbf{x}'} m_{\mathbf{x}} \frac{1}{-\nabla^2_{\mathbf{x}\mathbf{x}'}} m_{\mathbf{x}'}}. \tag{5.6}$$

The Gaussian contribution Z_0 describes small fluctuations of the phases around the ground state configuration, called spin waves. The topological excitations form a neutral 2D Coulomb gas in the $N = 0$ superselection sector [47].

So we have established that the model admits topological vortex configurations. But do these actually contribute to the correlation functions? And if yes, to what extent? To answer these questions we first note that the Coulomb potential in 2D is logarithmic, see (2.19). Therefore, the energy of a single vortex is $E = \pi J\beta \ln(L/\ell)$, where L is the linear dimension of the system, playing the role of an infrared cutoff and, in the present setting, we have chosen the lattice spacing $\ell = 1$. This vortex can be placed in any one of the plaquettes of the lattice of volume L^2. Therefore there are $\exp(2\ln L)$ different possible locations and the entropy of the vortex is thus also logarithmic, $S = 2\ln L$. The vortex can be assigned a free energy

$$F = (\pi J\beta - 2) \ln L. \tag{5.7}$$

Vortices proliferate if the free energy is dominated by the entropy, i.e. if $T > T_{\text{cr}} = \pi J/2$. Below this critical temperature they are suppressed by their large energy. This is the famed Berezinskii–Kosterlitz–Thouless (BKT) vortex unbinding phase transition [14].

To compute spin correlation functions

$$g\left(\mathbf{x}_1, \mathbf{x}_2\right) = \left\langle e^{i\left(\varphi_{\mathbf{x}_2} - \varphi_{\mathbf{x}_1}\right)} \right\rangle, \tag{5.8}$$

we couple the model to $\exp i \sum_{\mathbf{x},i} \left(\Delta_i \varphi_{\mathbf{x}} - 2\pi n_{\mathbf{x},i}\right) l_{\mathbf{x},i}$ where the new variables $l_{\mathbf{x},i}$ take the value 1 on the links $\{\mathbf{x}, i\}$ of a path joining $\mathbf{x}_1$ to $\mathbf{x}_2$ and vanish everywhere else. We have also added the factor 1 in the form $\exp\left(-i 2\pi \sum_{\mathbf{x},i} n_{\mathbf{x},i} l_{\mathbf{x},i}\right)$. We now use the same decomposition (5.5) to obtain

$$g\left(\mathbf{x}_1, \mathbf{x}_2\right) = g_0\left(\mathbf{x}_1, \mathbf{x}_2\right) \cdot g_{\text{top}}\left(\mathbf{x}_1, \mathbf{x}_2\right),$$

$$g_0\left(\mathbf{x}_1, \mathbf{x}_2\right) = \frac{1}{Z_0} \int_{-\infty}^{+\infty} \mathcal{D}\varphi\, e^{-\frac{\beta J}{2} \sum_{\mathbf{x},i} (\Delta_i \varphi_{\mathbf{x}})^2} e^{i \sum_{\mathbf{x}} \xi_{\mathbf{x}} \varphi_{\mathbf{x}}} = \left|\mathbf{x}_1 - \mathbf{x}_2\right|^{-\frac{2\pi J}{T}},$$

$$g_{\text{top}}\left(\mathbf{x}_1, \mathbf{x}_2\right) = \frac{1}{Z_{\text{top}}} \sum_{\{m_{\mathbf{x}}\}} e^{-2\pi^2 J\beta \sum_{\mathbf{x}\mathbf{x}'} m_{\mathbf{x}} \frac{1}{-\nabla^2_{\mathbf{x}\mathbf{x}'}} m_{\mathbf{x}'}} e^{i \sum_{\mathbf{x}} \eta_{\mathbf{x}} m_{\mathbf{x}}},$$

$$\tag{5.9}$$

where

$$\eta_{\mathbf{x}} = 2\pi \sum_{i,j} \epsilon_{ij} \frac{\hat{\Delta}_i}{\nabla^2} l_{\mathbf{x},j} \tag{5.10}$$

represents a line of orthogonal dipoles along the path.

To proceed in the computation of the correlation function we introduce the dual representation of the Coulomb gas, by expressing g_{top} as

$$g_{\text{top}}\left(\mathbf{x}_1, \mathbf{x}_2\right) = \frac{1}{Z_{\text{top}}} \int_{-\pi}^{+\pi} \mathcal{D}\chi\, e^{-\sum_{\mathbf{x},i} \frac{1}{8\pi^2 J\beta} (\Delta_i \chi_{\mathbf{x}})^2}$$

$$\times \sum_N \frac{z^N}{N!} \sum_{x_1,\ldots,x_N} \sum_{m_1,\ldots,m_n = \pm 1} e^{i \sum_{\mathbf{x}} m_{\mathbf{x}} (\chi_{\mathbf{x}} + \eta_{\mathbf{x}})}, \tag{5.11}$$

where

$$z = e^{-2\pi^2 J\beta G(0)} \tag{5.12}$$

is the vortex fugacity and we have adopted the dilute vortex approximation in which we consider only fundamental vortices $m_{\mathbf{x}} = \pm 1$. $G(0)$ is the infrared-regularized value of the lattice Coulomb kernel at coinciding points. The sums can be now computed, with the result

$$g_{\text{top}}\left(\mathbf{x}_1, \mathbf{x}_2\right) = \frac{1}{Z_{\text{top}}} \int_{-\infty}^{+\infty} \mathcal{D}\chi\, e^{-\sum_{\mathbf{x},i} \frac{1}{8\pi^2 J\beta}(\Delta_i \chi_{\mathbf{x}})^2 + 2z(1-\cos(\chi_{\mathbf{x}}+\eta_{\mathbf{x}}))}.$$

(5.13)

By shifting $\chi_{\mathbf{x}}$ by $-\eta_{\mathbf{x}}$ and introducing $\mu^2 = 8\pi^2 J\beta z$ we can rewrite this as

$$g_{\text{top}}\left(\mathbf{x}_1, \mathbf{x}_2\right) = \frac{1}{Z_{\text{top}}} \int_{-\infty}^{+\infty} \mathcal{D}\chi\, e^{-\frac{1}{4\pi^2 J\beta} \sum_{\mathbf{x},i} \frac{1}{2}(\Delta_i(\chi_{\mathbf{x}}-\eta_{\mathbf{x}}))^2 + \mu^2(1-\cos(\chi_{\mathbf{x}}))}.$$

(5.14)

For $T \gg T_{\text{cr}}$, this integral is dominated by the classical solution to the equation of motion

$$\nabla^2 \chi_{\mathbf{x}}^{\text{cl}} = \nabla^2 \eta_{\mathbf{x}} + \mu^2 \sin\chi_{\mathbf{x}}^{\text{cl}}.$$

(5.15)

For simplicity of presentation, let us assume that the path joining $\mathbf{x}_1$ to $\mathbf{x}_2$ lies on the y axis. Far from the endpoints of this path, where $l_{\mathbf{x},i} \neq 0$, the saddle point equation reduces to a one-dimensional equation.

$$\hat{\Delta}_1 \Delta_1 \chi_{\mathbf{x}}^{\text{cl}} = 2\pi \hat{\Delta}_1 l_{\mathbf{x},2} + \mu^2 \sin\chi_{\mathbf{x}}^{\text{cl}}.$$

(5.16)

Following [19] we solve this equation in the continuum limit,

$$\partial_x \partial_x \chi^{\text{cl}} = 2\pi \delta'(x) + \mu^2 \sin\chi^{\text{cl}}.$$

(5.17)

The solution with the boundary conditions $\chi^{\text{cl}} \to 0$ for $|x| \to \infty$ is

$$\chi^{\text{cl}} = \text{sign}(x)\, 4 \arctan e^{-\mu|x|}.$$

(5.18)

Inserting this back in (5.14) we get

$$g_{\text{top}}\left(\mathbf{x}_1, \mathbf{x}_2\right) = e^{-\frac{|\mathbf{x}_1 - \mathbf{x}_2|}{\xi_{XY}}},$$

$$\xi_{XY} = \sqrt{\frac{\pi^2 J}{8 T z}}.$$

(5.19)

The vortex unbinding BKT transition is thus accompanied by a change in behaviour of the correlation function. Above the transition, where vortices are free, this decays exponentially, with a correlation length ξ_{XY}. When the fugacity $z \to 0$ at the transition the correlation length diverges and the correlation functions become power laws, see (5.9). These algebraic correlation functions reflect the impossibility of long-range order in 2D at finite temperatures, as implied by the Hohenberg–Mermin–Wagner theorem [58].

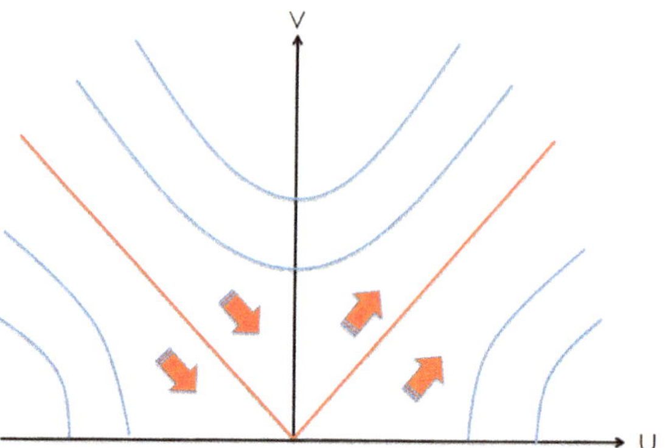

Fig. 5.1. The renormalization flow of the sine–Gordon model. Arrows indicate the flow towards the infrared regime.

Only at $T = 0$ the exponent vanishes, the correlation function becomes a constant and there is true long-range order.

The divergence of the correlation length at the critical temperature is singular [14],

$$\xi_{XY} \propto e^{\sqrt{\frac{\xi^+}{|T - T_{\mathrm{cr}}|}}}, \quad T \to T_{\mathrm{cr}}^+. \tag{5.20}$$

This means that, strictly speaking, the BKT transition is an infinite-order transition and follows from the observation that the dual formulation of the 2D Coulomb gas in (5.13), with partition function

$$Z_{\mathrm{top}} = \int_{-\infty}^{+\infty} \mathcal{D}\chi \, e^{- \sum_{\mathbf{x},i} \frac{1}{8\pi^2 J\beta} (\Delta_i \chi_{\mathbf{x}})^2 + 2z(1 - \cos\chi_{\mathbf{x}})}, \tag{5.21}$$

is a well known field-theoretic model, called the sine–Gordon model (for a review see [59]). The renormalization flow of the sine–Gordon model is expressed best in terms of the two variables

$$\begin{aligned} u &= 1 - \frac{T_{\mathrm{cr}}}{T}, \\ v &= 16\pi z \frac{T_{\mathrm{cr}}}{T}, \end{aligned} \tag{5.22}$$

and is shown schematically in Fig. 5.1.

There is a line of $z = 0$ infrared fixed points for $T < T_{\mathrm{cr}}$, all corresponding to states with bound vortices and differing by a constant representing the initial conditions of the flow equations. These infrared fixed points describe 2D superfluid states or, when the model is coupled to electromagnetic fields, 2D superconductors [47].

5.2. Quantum wires and phase slips

We have seen a first example in which topological excitations can have a large effect on the classical ground state of a model. Let us now consider exactly the same model (5.4) as the Euclidean partition function of a 1D quantum model, where we substitute βJ with a dimensionless quantum coupling g. It turns out that this model describes the quantum phase fluctuations on a Josephson junction chain [7] with completely screened Coulomb interactions, or, in its continuum formulation, the dynamics of a 1D quantum wire [60], a wire in which quantum effects dominate charge transport. The quantum coupling g encodes the dimensionless conductance of the system. We will focus on the details of such systems in a later chapter. Here, suffice to say that, for high g, charge transport by quantum tunneling dominates the self-energy of the charge carriers while for low g the opposite is true.

In this case, one of the two directions of the 2D lattice is the Euclidean time. Therefore, the 2π circulation of the phase around the boundary of a plaquette that is used to characterize a vortex at its centre becomes an event, rather than a static topological excitation. This event is physically realized when the phase on one site of the chain flips by 2π in a given characteristic time, representing the lattice spacing in the Euclidean time direction, or when two adjacent phases both simultaneously flip by π in the same sense. This quantum tunneling between two equivalent configurations of the phase is called a *phase slip* [61–63] and is the first example of *instanton* we will discuss.

From our discussion of vortices we immediately know that phase slip instantons proliferate when the conductance g is low enough. But do they also have large effects on the quantum ground state? To answer this question we first note that $\Delta_\mu \varphi$, being canonically conjugate to the current, plays the role of a gauge field in the quantum wire model, where now the index $\mu = 0, 1$. We can then use the standard tool to probe the behaviour of a gauge theory, namely the Wilson loop operator (for a review see [19,20].) The Wilson loop is the phase factor given by the circulation of the gauge field on a closed loop C. Suppose we are given a generic gauge theory with action $S(A_\mu)$ for gauge fields A_μ. When the loop C is chosen as a rectangle of dimensions T and L in the plane formed by the Euclidean time direction and one spatial direction, the Wilson loop measures the amplitude for separating two probe charges $\pm q_{\rm ext}$ a distance L apart, letting them propagate for a time T and then bringing them together again. When T is very large, only the ground state contributes to this amplitude and

$$\langle W(C)\rangle = \frac{1}{Z_A} \int \mathcal{D}A_\mu e^{-S(A_\mu)} e^{iq_{\rm ext} \oint_C A_\mu d^\mu x} = e^{-TV_{q_{\rm ext}}(L)}, \qquad (5.23)$$

measures the ground state potential $V_{q_{\rm ext}}(L)$ between the two probe charges. Therefore, a perimeter law for the Wilson loop indicates a short-range potential, while an *area law* is tantamount to a linear potential between probe charges. A linear potential $V(L) = \sigma L$ implies *confinement* of charges, since an infinite energy is required to completely separate them. Note, however, that this is different from the logarithmic confinement of vortices in the BKT transition [14] discussed previously. Contrary to logarithmic confinement, linear confinement is characterized by a new length scale $\sqrt{\hbar c/\sigma}$ given by the inverse square root of the *string tension* σ. This is the typical length of a *string* binding the charges together. In a model with dynamical charges, when the string is pulled apart on distances larger than this characteristic length scale it becomes energetically favourable to pop out of the vacuum additional charges of both signs and form several short strings instead of a single long one.

In the quantum wire model we have

$$\langle W(C)\rangle = \frac{1}{Z} \sum_{\{n_x\}} \int_{-\pi}^{+\pi} \mathcal{D}\varphi \, e^{-\frac{g}{2}\sum_{x,\mu}(\Delta_\mu \varphi_x - 2\pi n_{x,\mu})^2}$$

$$\times e^{iq_{\rm ext}\sum_C l_{x,\mu}(\Delta_\mu \varphi_x - 2\pi n_{x,\mu})} \qquad (5.24)$$

where $l_{x,\mu} = 1$ on the links forming the loop C and $l_{x,\mu} = 0$ everywhere else and we have indicated explicitly the sum over Euclidean space-time indices. Also, we have inserted 1 in the form $\exp(-i2\pi q_{\rm ext}\sum_C l_{x,\mu}n_{x,\mu})$. We can now use the lattice Stokes theorem by representing

$$l_{x,\mu} = \epsilon_{\mu\nu}\Delta_\nu S_x, \qquad (5.25)$$

where the surface element $S_x = 1$ on the plaquettes comprising the surface S encircled by the loop C and $S_x = 0$ everywhere else. Finally we repeat all steps that led to (5.6) to obtain

$$\langle W(C)\rangle = \frac{1}{Z_{\rm top}} \sum_{\{m_x\}} e^{-2\pi^2 g \sum_{xx'} m_x \frac{1}{-\nabla^2_{xx'}} m_{x'}} e^{i2\pi q_{\rm ext}\sum_S S_x m_x}. \qquad (5.26)$$

It is now evident that phase slip instantons have no dramatic effect on fundamental charges $q_{\text{ext}} = 1$ (Cooper pairs in the quantum wire model). They have been shown to cause the transition from a superconductor to a metal by introducing energy dissipation [61] or even to an insulator in finite systems coupled to the environment [64], but they do not fundamentally alter the quantum state of charges. This is not so for half-charges $q_{\text{ext}} = 1/2$ (single electrons in the quantum wire model). If we set $q_{\text{ext}} = 1/2$ in (5.26) we get an area law behaviour for the Wilson loop. This is in concert with the well-known result that instantons confine half-charges in the 1D Abelian Higgs model [65].

5.3. Compact QED in 2D

Let us now focus on a model in which instantons do have a dramatic effect on fundamental charges. Instead of a model with a global O(2) symmetry in 1D we shall consider a compact gauge theory in 2D. By absorbing a factor ℓ into its definition, we consider a dimensionless gauge field A_μ living on the links of a 2D square lattice of spacing ℓ (we omit the lattice vertex subscript from now on to simplify notation), with action invariant under Abelian gauge transformations $A_\mu \to A_\mu + \Delta_\mu \lambda$. If A_μ is a real variable and the Euclidean action has the usual Maxwell form

$$S = \frac{1}{4e^2} \sum_x F_{\mu\nu} F_{\mu\nu} = \frac{1}{2e^2} \sum_x F_\mu F_\mu, \tag{5.27}$$

where $F_\mu = K_{\mu\nu} A_\nu$ is the dual field strength, then the gauge group is $\mathbb{R}$, which is a non-compact group. If, instead, the gauge field is an angular variable, $A_\mu \in [-\pi, +\pi]$, then the gauge group is compact U(1) and the action must be chosen to reflect the periodicity [18, 19],

$$S = \frac{1}{e^2} \sum_{x,\mu} (1 - \cos(F_\mu)). \tag{5.28}$$

As already pointed out in Chapter 3, in 2D the coupling constant has canonical dimension [mass]. The dimensionless coupling e^2 appearing in (5.27) and (5.28) is the product of such coupling constant with the lattice spacing.

In the following we shall again consider the Villain formulation of the compact model,

$$Z = \sum_{\{n_\mu\}} \int_{-\pi}^{+\pi} \mathcal{D}A_\mu \, e^{-\frac{1}{2e^2} \sum_{x,\mu} (F_\mu - 2\pi n_\mu)^2}. \tag{5.29}$$

The integers $2\pi n_\mu$ represent, as in the quantum wire model, saddle point configurations. The dual field strength F_μ is peaked around these configurations for weak enough coupling e^2. In general, since the epsilon tensor is antisymmetric, the dual field strength satisfies $\hat{\Delta}_\mu F_\mu = 0$. Due to the compactness of the gauge symmetry, however, the saddle points do not have to satisfy this equation. The integers $m = \hat{\Delta}_\mu n_\mu$ are the instantons of the model. From the Euclidean 3D point of view, the vector F_μ represents a magnetic field. These localized instantons are thus *magnetic monopoles*, assigned to the centres of lattice cubes, exactly as the quantum wire vortex instantons were assigned to plaquettes. In this case, the Dirac string represents the "time evolution" of a vortex on the 2D lattice. Suppose that it lies on the Euclidean time direction, encoding a static vortex on the physical 2D array. A magnetic monopole instanton in the centre of the 3D Euclidean cube above this plaquette represents then an event in which the flux of the original vortex is spread on all facettes of the cube. As we will see in a later chapter, this becomes particularly manifest in the deeply non-relativistic version of the monopole instantons, which are tunneling events between quantum states differing by one vortex number and represent therefore quantum fluctuations in which vortices appear and disappear from the vacuum. These events are the 2D analogue of quantum phase slips.

To proceed we use again the general decomposition of a set of integers $\{n_\mu\}$ [19]

$$n_\mu = K_{\mu\nu} k_\nu + K_{\mu\nu} \xi_\nu + \Delta_\mu \lambda, \tag{5.30}$$

where $\{k_\mu\}$ are integers and $\hat{\Delta}_\mu \Delta_\mu \lambda = \nabla^2 \lambda = m$. As before we use the integers $\{k_\mu\}$ to shift the domain of the gauge fields from $[-\pi, +\pi]$ to $[-\infty, +\infty]$ and we absorb the real variables $\{\xi_\mu\}$ into these now non-compact gauge fields. We have thus replaced a set of three integers $\{n_\mu\}$ with two integers $\{k_\mu\}$ (there are only two because of gauge invariance under transformations $k_\mu \to k_\mu + \Delta_\mu \chi$) and one integer $\{m\}$ representing the instantons. This gives

$$Z = Z_0 \cdot Z_{\text{inst.}} = \int_{-\infty}^{+\infty} \mathcal{D}A_\mu \, e^{-\frac{1}{2e^2} \sum_{x,\mu} F_\mu^2} \cdot \sum_{\{m\}} e^{-\frac{2\pi^2}{e^2} \sum_x m \frac{1}{-\nabla^2} m}. \tag{5.31}$$

This is the product of the partition function of the non-compact gauge theory with an additional contribution from the magnetic monopole instantons. Differently from the quantum wire model, the Coulomb interaction of the instantons is a 3D one, and thus behaves like $1/|x|$ rather than a logarithm.

As a consequence, the monopole instantons are always in a plasma phase, independently of the strength of the coupling.

To establish the consequence of this monopole plasma we measure again the Wilson loop expectation value by coupling the model to $\exp i q_{\text{ext}} \sum_C l_\mu A_\mu$ where $l_\mu = 1$ on the links forming the closed loop C and $l_\mu = 0$ everywhere else and the integer q_{ext} denotes the value of the external probe charges,

$$\langle W(C) \rangle = \frac{1}{Z} \sum_{\{n_\mu\}} \int_{-\pi}^{+\pi} \mathcal{D}A_\mu \, e^{-\frac{1}{2e^2} \sum_{x,\mu} (F_\mu - 2\pi n_\mu)^2} e^{i q_{\text{ext}} \sum_C l_\mu A_\mu}. \quad (5.32)$$

We now use the lattice Stokes theorem by representing $l_\mu = \hat{K}_{\mu\nu} S_\nu$, where $S_\mu = 1$ on the plaquettes spanning the surface S encircled by the loop C and $S_\mu = 0$ everywhere else,

$$\langle W(C) \rangle = \frac{1}{Z} \sum_{\{n_\mu\}} \int_{-\pi}^{+\pi} \mathcal{D}A_\mu \, e^{-\frac{1}{2e^2} \sum_{x,\mu} (F_\mu - 2\pi n_\mu)^2} e^{i q_{\text{ext}} \sum_S S_\mu (F_\mu - 2\pi n_\mu)},$$
$$(5.33)$$

where we have added a factor 1 in the form $\exp\left(-i2\pi q_{\text{ext}} \sum_S S_\mu n_\mu\right)$. Following the same steps leading to (5.31) we obtain

$$\langle W(C) \rangle = \frac{1}{Z_{\text{inst.}}} \sum_{\{m\}} e^{-\frac{2\pi^2}{e^2} \sum_x m \frac{1}{-\nabla^2} m} e^{i2\pi q_{\text{ext}} \sum_S mn}, \quad (5.34)$$

where

$$\eta = \frac{\hat{\Delta}_\mu}{\nabla^2} S_\mu \quad (5.35)$$

represents a unit dipole sheet on the surface S enclosed by the Wilson loop C. To compute the instanton contribution to the Wilson loop we introduce again the sine–Gordon dual representation of the monopole plasma,

$$\langle W(C) \rangle = \frac{1}{Z_{\text{inst.}}} \int_{-\pi}^{+\pi} \mathcal{D}\chi \, e^{-\sum_{x,\mu} \frac{e^2}{8\pi^2} (\Delta_\mu \chi)^2}$$
$$\times \sum_N \frac{z^N}{N!} \sum_{x_1,\ldots,x_N} \sum_{m_1,\ldots,m_n = \pm 1} e^{i \sum_x m(\chi + \eta)}, \quad (5.36)$$

where

$$z = e^{-\frac{2\pi^2}{e^2} G(0)} \quad (5.37)$$

is the monopole fugacity and $G(0)$ is the value of the 3D lattice Coulomb kernel at coinciding points. For e^2 not too large, monopoles of high charge tend to dissociate into unit charge monopoles [19]. We can then adopt the dilute monopole approximation, in which we consider only fundamental monopoles $m = \pm 1$, thereby obtaining

$$\langle W(C) \rangle = \frac{1}{Z_{\text{inst.}}} \int_{-\infty}^{+\infty} \mathcal{D}\chi \, e^{-\sum_{x,\mu} \frac{e^2}{8\pi^2}(\Delta_\mu \chi)^2 + 2z(1-\cos(\chi+\eta))}. \tag{5.38}$$

By shifting χ by $-\eta$ and introducing $\mu^2 = \frac{8\pi^2}{e^2} z$ we can rewrite this as

$$\langle W(C) \rangle = \frac{1}{Z_{\text{inst.}}} \int_{-\infty}^{+\infty} \mathcal{D}\chi \, e^{-\frac{e^2}{4\pi^2} \sum_{x,\mu} \frac{1}{2}(\Delta_\mu(\chi-\eta))^2 + \mu^2(1-\cos(\chi))}. \tag{5.39}$$

We shall evaluate this integral by the saddle point approximation. To this end we derive the classical equation of motion

$$\nabla^2 \chi^{\text{cl}} = \nabla^2 \eta + \mu^2 \sin \chi^{\text{cl}}. \tag{5.40}$$

Let us assume that the Wilson loop lies on the plane spanned by the Euclidean time direction and the y-axis. Far from the boundary of the enclosed surface S, where $S_\mu \neq 0$, the saddle point equation reduces to a one-dimensional equation in the x variable,

$$\hat{\Delta}_1 \Delta_1 \chi^{\text{cl}} = 2\pi q_{\text{ext}} \hat{\Delta}_1 S_1 + \mu^2 \sin \chi^{\text{cl}}. \tag{5.41}$$

Following [19] we solve this equation in the continuum limit,

$$\partial_x \partial_x \chi^{\text{cl}} = 2\pi q_{\text{ext}} \delta'(x) + \mu^2 \sin \chi^{\text{cl}}. \tag{5.42}$$

For fundamental charge probes $q_{\text{ext}} = 1$, the solution with the boundary conditions $\chi^{\text{cl}} \to 0$ for $|x| \to \infty$ is

$$\chi^{\text{cl}} = \text{sign}(x) \, 4 \arctan e^{-\mu|x|}. \tag{5.43}$$

Finally, inserting this back in (5.39) we get

$$\langle W(C) \rangle = e^{-\sigma A},$$

$$\sigma = \sqrt{\frac{8e^2 z}{\pi^2}} \frac{1}{\ell^2}, \tag{5.44}$$

where ℓ is the lattice spacing and A the area enclosed by the Wilson loop.

Contrary to phase slips in quantum wires, magnetic monopole instantons in 2D lead to an area law for fundamental unit charges $q_{\text{ext}} = \pm 1$. These are thus confined by a linear potential with string tension σ due to the formation of an electric flux tube between the probe charges. In the following chapters we will explore what are the consequences in real materials. Before doing that, however, let us note that also half-charges are confined by monopoles. For $q_{\text{ext}} = 1/2$, the solution to (5.42) with vanishing boundary conditions is

$$\chi^{\text{cl}} = \Theta(x)\, 4\arctan \mathrm{e}^{-\mu|x|}, \tag{5.45}$$

implying a string tension $\sigma_{1/2} = (1/2)\sigma$.

Let us now compute correlation functions of gauge invariant (dual) field strengths. These decompose into

$$\langle F_\mu(x) F_\nu(y) \rangle_C = \langle F_\mu(x) F_\nu(y) \rangle_C^0 + \langle F_\mu(x) F_\nu(y) \rangle_C^{\text{mon}}, \tag{5.46}$$

where

$$\langle F_\mu(x) F_\nu(y) \rangle_C^0 = e^2 \left(\delta_{\mu\nu} - \frac{\Delta_\mu \hat{\Delta}_\nu}{\nabla^2} \right) \delta_{x,y} \tag{5.47}$$

is the usual contribution from massless photons and the second term is the modification due to magnetic monopole instantons. To compute it, we first note that, as evident from (5.33), for $q_{\text{ext}} = 1$, we can replace the monopole contribution to F_μ by the operator $2\pi\delta/i\delta S_\mu$ acting on the Wilson loop. We then keep only quadratic contributions to the cosine in (5.38) and we integrate over the field χ. This gives the following representation of the Wilson loop,

$$\langle W(C) \rangle = \mathrm{e}^{-W+\cdots},$$

$$W = \sum_x \frac{e^2}{4\pi^2} S_\mu \frac{\mu^2 \Delta_\mu \hat{\Delta}_\nu}{\nabla^2(-\nabla^2 + \mu^2)} S_\nu, \tag{5.48}$$

which, combined with (5.47) gives the final result

$$\langle F_\mu(x) F_\nu(y) \rangle_C = e^2 \left(\delta_{\mu\nu} + \frac{\Delta_\mu \hat{\Delta}_\nu}{(-\nabla^2 + \mu^2)} \right) \delta_{x,y}. \tag{5.49}$$

This shows that the monopole plasma induces a mass μ for the photon, although there is no explicit gauge symmetry breaking in the action.

There are thus two scales in the problem. The photon mass μ determines the width $w = 1/\mu$ of the electric flux tubes, while the string tension sets their typical length $d = 1/\sqrt{\sigma}$, both in units of the lattice spacing. The ratio of these two scales behaves as

$$\frac{w}{d} \propto \frac{e^{3/2}}{z^{1/4}}.$$ (5.50)

This behaviour, characteristic of Abrikosov-type flux tubes, has to be contrasted with the quantum chromodynamics Nambu–Goto strings, for which the dependence of the width-to-length ratio on the coupling constant is a very mild one [66].

Chapter 6

Effective Chern–Simons gauge theories of emergent condensed matter systems

The Landau theory of second-order phase transitions (for a review see [67]) is one of the pillars of modern statistical mechanics and condensed matter physics. It uses an order parameter spontaneously breaking a given symmetry to distinguish between an ordered and a disordered phase. There are, however, quantum phase transitions (for a review see [68]) between phases that cannot be distinguished by different symmetries, the most notable examples being the transitions between different fractional quantum Hall plateaus at different filling factors (for a review see [69]).

Spin-polarized, planar electrons of mass m and electric charge e, subject to an external, uniform magnetic field $B > 0$, are governed by the one-particle Hamiltonian

$$H = -\frac{1}{2m}\left(\nabla - ie\mathbf{A}\right)^2 - \frac{eB}{2m}, \tag{6.1}$$

where the second term represents the Pauli interaction. Focusing on circular geometries, we choose the symmetric gauge $A_i = (B/2)\epsilon^{ij}x^j$, $i,j = 1, 2$, for the external vector potential. The fundamental scale set by the external magnetic field is the magnetic length, $\ell = \sqrt{2/eB}$. Introducing the complex notation $z = x^1 + ix^2$, $\bar{z} = x^1 - ix^2$, $\partial = \partial/\partial z$, $\bar{\partial} = \partial/\partial \bar{z}$, and two commuting sets of harmonic oscillator operators

$$d = \frac{z}{2\ell} + \ell\bar{\partial}, \quad d^\dagger = \frac{\bar{z}}{2\ell} - \ell\partial, \quad [d, d^\dagger] = 1,$$

$$c = \frac{\bar{z}}{2\ell} + \ell\partial, \quad c^\dagger = \frac{z}{2\ell} - \ell\bar{\partial}, \quad [c, c^\dagger] = 1, \tag{6.2}$$

the Hamiltonian (6.1) and the canonical angular momentum $J = -ix^i \varepsilon^{ij} \partial_j$ can be rewritten as

$$H = \omega\, d^\dagger d,$$
$$J = c^\dagger c - d^\dagger d, \qquad\qquad (6.3)$$

where $\omega = eB/m$ is the cyclotron frequency. Since the operators c and d commute, the spectrum consists of infinitely degenerate levels of energy $\varepsilon_n = \omega n$: these are called the Landau levels. The degenerate states in one Landau level are characterized by the angular momentum eigenvalue l. Wave functions ψ of the first Landau level satisfy $d\psi = 0$ and have vanishing energy due to the additional Pauli interaction in (6.1). A complete basis is given by

$$\psi_l(z, \bar z) = \frac{1}{\ell\sqrt{\pi}}\, \frac{1}{\sqrt{l!}}\, \left(\frac{z}{\ell}\right)^l\, e^{-|z|^2/2\ell^2}. \qquad\qquad (6.4)$$

Note that the radial part of these wavefunctions is sharply peaked around a radius $r_l = \ell\sqrt{l}$. Higher Landau levels' basis states can be obtained by repeated action of the operators $d^\dagger$ and $c^\dagger$ on ψ_0:

$$\begin{aligned}
\psi_{n,l}(z, \bar z) &= \frac{(c^\dagger)^{l+n}}{\sqrt{(l+n)!}}\, \frac{(d^\dagger)^n}{\sqrt{n!}} \psi_0(z, \bar z) \\[2mm]
&= \frac{1}{\ell}\, \sqrt{\frac{n!}{\pi(l+n)!}}\, \left(\frac{z}{\ell}\right)^l\, L_n^l\left(\frac{|z|^2}{\ell^2}\right)\, e^{-|z|^2/2\ell^2}.
\end{aligned} \qquad (6.5)$$

Here $n = 0, 1, \dots$ labels the energy level, $l + n \geq 0$, and $L_n^l(x)$ are the generalized Laguerre polynomials.

The many-body problem of N electrons is described, in second quantization, by the Hamiltonian

$$H = \frac{1}{2m} \int d^2\mathbf{x}\, (D_i\Psi)^\dagger (D_i\Psi) - \frac{1}{2m} \int d^2\mathbf{x}\, eB\rho, \qquad (6.6)$$

with $D_i = \partial_i + ieA_i$ the covariant derivative and $\rho(\mathbf{x}, t) = \Psi^\dagger(\mathbf{x}, t)\Psi(\mathbf{x}, t)$ the particle-number density. The field operator $\Psi(\mathbf{x}, t)$ possesses an expansion in terms of the single-particle energy and angular momentum eigenstates $\psi_{n,l}(\mathbf{x})$,

$$\Psi(\mathbf{x}, t) = \sum_{n=0}^{\infty} \sum_{l=-n}^{\infty} a_l^{(n)} \psi_{n,l}(\mathbf{x})\, e^{-in\omega t}. \qquad (6.7)$$

The coefficients are fermionic Fock annihilators and creators satisfying

$$\{a_k^{(n)}, a_l^{(m)\dagger}\} = \delta_{n,m}\delta_{k,l}, \tag{6.8}$$

with all other anticommutators vanishing. In the following we will consider only the physics of the first Landau level, for which the field operator takes the (time-independent) form

$$\Psi(\mathbf{x}) = \sum_{l=0}^{\infty} a_l\, \psi_l(\mathbf{x}), \tag{6.9}$$

with $\psi_l(\mathbf{x})$ given by (6.4) and $a_l \equiv a_l^{(0)}$. The property $d\psi_l = 0$ can be also written as a self-dual condition on the field operator

$$D_+\Psi(\mathbf{x}) \equiv (D_1 + iD_2)\,\Psi(\mathbf{x}) = 0. \tag{6.10}$$

Let us consider a many-body state consisting of one completely filled Landau level on a disk geometry. By this we mean a state $|\Omega\rangle$ in which all angular momentum states of the first Landau level are occupied up to (and including) a maximal angular momentum L (the total number of particles is then $N = L + 1$):

$$|\Omega\rangle = a_0^\dagger a_1^\dagger \ldots a_L^\dagger |0\rangle, \tag{6.11}$$

where $|0\rangle$ is the Fock vacuum. Since the single-particle angular momentum states are peaked around radii $r_l = \ell\sqrt{l}$, this state consists of a circular droplet of radius approximately given by $R \simeq \ell\sqrt{L}$. This configuration is clearly *incompressible*: a compression of the droplet would lower its total angular momentum and face an energy gap ω, since at least one electron would be promoted to the next Landau level. The incompressibility is reflected in the spatial distribution of the matter. The expectation value of the density is easy to compute in the second-quantized formalism,

$$\langle\Omega|\rho(\mathbf{x})|\Omega\rangle = \frac{1}{\ell^2\pi}\, e^{-r^2/\ell^2} \sum_{l=0}^{L} \frac{1}{l!} \left(\frac{r}{\ell}\right)^{2l}. \tag{6.12}$$

It is constant for $r \ll \ell\sqrt{L}$ and drops rapidly to zero around $r \simeq \ell\sqrt{L}$. Thus, incompressibility implies that the fully filled first Landau level forms a droplet of uniform density. Another quantity of interest is the expectation value of the current

$$J^i(\mathbf{x}) = \frac{1}{2im} \left\{ \Psi^\dagger(\mathbf{x})D_i\Psi(\mathbf{x}) - (D_i\Psi(\mathbf{x}))^\dagger\Psi(\mathbf{x}) \right\}. \tag{6.13}$$

This is easily obtained by combining (6.13) with $\partial_i \rho = \Psi^\dagger D_i \Psi + (D_i \Psi)^\dagger \Psi$ and using the self-dual condition (6.10), which gives

$$\langle \Omega | J^i(\mathbf{x}) | \Omega \rangle = -\frac{1}{2m} \varepsilon^{ij} \partial_j \langle \Omega | \rho(\mathbf{x}) | \Omega \rangle. \tag{6.14}$$

The current vanishes in the interior of the droplet, where the density is constant. However, the state supports a transverse, circular current around the boundary of the droplet. This shows that the only dynamics of an incompressible quantum Hall droplet is concentrated around the boundary.

Suppose now that an in-plane electric field is applied. Each electron in the droplet experiences the Lorentz force. If electrons would be independent of each other, each of them would move on a curved trajectory between scatterings with impurities and phonons. Since they are in a gapped multi-body ground state, however, uncorrelated movement would imply that some of the electrons are promoted to the next Landau level, which comes at an energy cost. Since the Lorentz force does no work, the energy is minimized by moving the incompressible droplet as a whole in a direction orthogonal to the applied electric field, which results in a Hall conductivity $\sigma_H = \nu e^2 / 2\pi$ with $\nu = 1$. The strong correlations between elementary constituents of the macroscopic droplet render impurities and phonons irrelevant. The macroscopic quantum state behaves as if these would not exist. This property is the reason why this integer quantum Hall effect permits one of the most accurate measurements of the fine structure constant to date, although it involves a macroscopic number of electrons [70].

Since the first measurement of the integer quantum Hall effect [70], other macroscopic, strongly correlated quantum states have been found, corresponding to fractional values of the filling fraction ν in the Hall conductivity [71], see Fig. 6.1. They all correspond to incompressible quantum fluids [72] but of interacting electrons in this fractional case.

The interacting incompressible quantum fluids are all characterized by a universal, very precise fractional Hall conductivity. However, they are not distinguished by any order parameter and, thus, they do not fall into the Landau scheme. Another principle must be at work. In general, in field theory, fractional quantum numbers are associated either with an infinite-dimensional dynamical symmetry, as in the example of conformal field theory [73], or with the topology of gauge fields [40]. Both avenues have been explored [74–76] and have led essentially to the same results. Here we shall focus on topological gauge theories, since they have led to the concept of topological order [77, 78] which will be crucial for what follows.

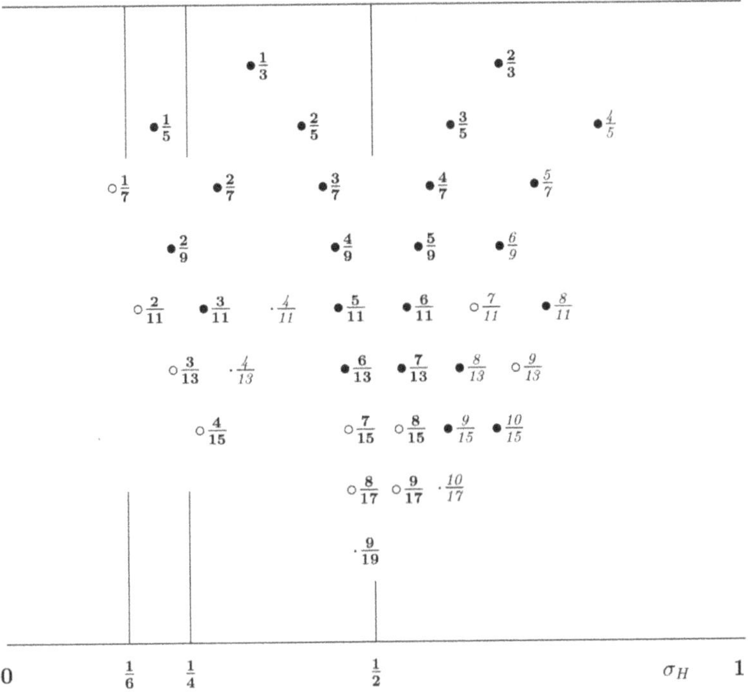

Fig. 6.1. Experimentally observed fractional Hall conductivity plateaus σ_H in units of e^2/h in the range $0 < \sigma_H < 1$, displayed in rows corresponding to their denominator. The thickness of the dots denote their stability: ($\bullet$) very stable, ($\circ$) stable, ($\cdot$) less stable. In **bold** the theoretically best understood plateaus, in *italics* the ones that fit less well established hierarchies. From A. Cappelli, C. A. Trugenberger and G. Zemba, *Nucl. Phys.* **448**, 470 (1995), ©Elsevier (1995).

The basic idea of the construction is simple [76]. Let us consider emergent condensed matter ground states in 2D. For the simplest cases, the excitations above these ground states give rise to a single conserved current j^μ describing density fluctuations. Any conserved vector current in 2D can be formulated as the curl of a pseudovector field a_μ as

$$j^\mu = \frac{1}{2\pi}\epsilon^{\mu\alpha\nu}\partial_\alpha a_\nu. \tag{6.15}$$

Since this definition is invariant under transformations $a_\mu \to a_\mu + \partial_\mu\lambda$, the pseudovector field is actually a gauge field. The effective actions for such states of matter can thus be formulated as gauge theories. In presence of a parity- and time-reversal-breaking external magnetic field, e.g., the infrared-dominant term in the effective action is the topological Chern–Simons term,

$$\mathcal{L}_{\text{eff}} = \frac{k}{4\pi} a_\mu \epsilon^{\mu\alpha\nu} \partial_\alpha a_\nu. \tag{6.16}$$

To establish the nature of this topological ground state, let us compute its electromagnetic response by coupling the conserved current to electromagnetism,

$$\mathcal{L} \to \mathcal{L}_{\text{eff}} + e A_\mu j^\mu = \frac{k}{4\pi} a_\mu \epsilon^{\mu\alpha\nu} \partial_\alpha a_\nu + \frac{e}{2\pi} A_\mu \epsilon^{\mu\alpha\nu} \partial_\alpha a_\nu, \tag{6.17}$$

and by integrating out the fictitious gauge field a_μ. This is best done by adding a regulator Maxwell term with coefficient $1/f^2$ and then removing the regulator by taking $f \to \infty$ in the result. Using the techniques of Chapter 3, this gives the effective action

$$S_{\text{em}} = -\frac{e^2}{4\pi k} \int d^3x \, A_\mu \epsilon^{\mu\alpha\nu} \partial_\alpha A_\nu. \tag{6.18}$$

The current induced by an external electric field $\mathbf{E}$ in this state of matter is thus

$$j^i_{\text{ind}} = \frac{\delta}{\delta A_i} S_{\text{em}} = \frac{e^2}{2\pi k} \epsilon^{ij} E^j. \tag{6.19}$$

There is no longitudinal current, parallel to the applied electric field, $\sigma_{\text{long}} = 0$. The sole response of the system is a transverse current with Hall conductivity

$$\sigma_H = \frac{e^2}{2\pi} \nu, \quad \nu = \frac{1}{k}, \tag{6.20}$$

the Chern–Simons coupling playing the role of the inverse filling fraction.

Chern–Simons gauge theories are thus candidates to describe the long-distance physics of Laughlin's incompressible quantum Hall fluids [69]. These, however, are all characterized by filling fractions which are inverse odd integers, as can be seen in the topmost left corner of Fig. 6.1. How can we derive this quantization condition in the present effective field theory context? To this end we must consider localized topological excitations. The nature of these can be easily recognized by integrating by parts in (6.17). This shows that the fictitious gauge field a_μ couples to quantized fluxes Φ^μ,

$$\mathcal{L}_{\text{eff}} = \frac{k}{4\pi} a_\mu \epsilon^{\mu\alpha\nu} \partial_\alpha a_\nu - e a_\mu \Phi^\mu + \frac{e}{2\pi} A_\mu \epsilon^{\mu\alpha\nu} \partial_\alpha a_\nu. \tag{6.21}$$

These fluxes are vortices due to the compact nature of the gauge fields and represent the gapped quasi-particle excitations of the incompressible quantum fluid. Repeating the Gaussian integration over a_μ we obtain

$$S_\Phi = \int d^3x \left(\frac{\pi}{k} \Phi_\mu \epsilon^{\mu\alpha\nu} \frac{\partial_\alpha}{\nabla^2} \Phi_\nu + \frac{e}{k} A_\mu \Phi^\mu \right). \tag{6.22}$$

Comparing with (3.25) we recognize the first term as the Gaussian linking number of the quasi-particle trajectories, encoding the phase acquired by one quasi-particle when it encircles a second one. Half of this phase is the fractional statistics factor for the exchange of quasi-particles [79]. The second factor shows, instead, that quasi-particles carry both magnetic flux and charge. Quasi-particles carrying n flux quanta are thus anyons [79] with fractional statistics θ and charge q given by

$$\theta = \frac{n^2}{k}, \quad q = \frac{n}{k} e. \tag{6.23}$$

Quantum Hall samples are made of physical electrons. Therefore, we should require the existence of one excitation with the quantum numbers of the electron in the above spectrum. The requirement of unit charge can be met by choosing integer $k \in \mathbb{Z}$, the electron being the excitation with $n = -k$. This excitation has statistical parameter $\theta = k$. In order to recover the required fermionic statistics we must then choose k as on *odd* integer, $k = 2p + 1$, $p \in \mathbb{Z}$. This shows that only odd-denominator filling fractions are admitted for incompressible quantum fluids made of electrons,

$$\nu = \frac{1}{2p + 1}, \tag{6.24}$$

as observed in experiments. In these ground states the electron appears as a composite of a unit charge $-e$ and $(2p + 1)$ elementary magnetic fluxes. The topological character of the effective field theory accounts both for the quantization of the Hall conductivity, reflected in the extreme precision of the fine structure constant measurements, and for the absolute irrelevance of disorder for the properties of these emergent, strongly correlated ground states. Disorder can affect only geometry-dependent effective field theory terms; when the effective field theory does not depend on the metric, disorder becomes irrelevant.

Single Chern–Simons gauge fields describe the original Laughlin fluids [72] of filling fraction $\nu = 1/(2p + 1)$, for p an integer. The composite incompressible fluids of the full hierarchy [69] in Fig. 6.1 involve N flavours of gauge fields a_μ^I, labelled by the integer vector $\mathbf{l}$ and coupled by a so-called K matrix,

$$\mathcal{L}_{\text{eff}} = \frac{1}{4\pi} K^{IJ} a_\mu^I \epsilon^{\mu\alpha\nu} \partial_\alpha a_\nu^J + a_\mu^I q^{\mu I}, \tag{6.25}$$

where $q^{\mu I}$ are the corresponding quasi-particle excitations. These represent the singularities of the fields $j^{\mu I} = (1/2\pi)\epsilon^{\mu\alpha\nu}\partial_\alpha a_\nu^I$ due to the compactness of the gauge fields a_μ^I. Both the statistics of the quasi-particles and their mutual statistics are encoded in the K matrix,

$$\theta_1 = \pi\, \mathbf{1}^{\mathrm{T}} K^{-1} \mathbf{1},$$
$$\theta_{12} = 2\pi\, \mathbf{l}_1{}^{\mathrm{T}} K^{-1} \mathbf{l}_2. \tag{6.26}$$

Requiring, as above, that there exists quasi-particle excitations with the quantum numbers of the electron, leads to the observed filling fractions in Fig. 6.1 [76]. Finally, it can be shown that the ground state of the effective Chern–Simons gauge theory quantized on a Riemann surface of genus g has degeneracy $(\det K)^g$. This non-trivial ground state degeneracy, reflecting the number of different incompressible quantum fluids existing on Riemann surfaces, goes under the name of *topological order* (for a review see [80]).

Topological order is a consequence of incompressibility on compact spaces. How is incompressibility reflected on open disk geometries? To answer this question we re-examine the simplest effective field theory (6.16) for the Laughlin fluids. On compact spaces the action is gauge invariant, on an open disk it is not, since the Lagrangian changes by a total derivative term under transformations $a_\mu \to a_\mu + \partial_\mu \lambda$. To restore full gauge invariance we must introduce additional degrees of freedom χ on the boundary that compensate this gauge variation by a transformation $\chi \to \chi + \lambda$. Let us choose the Weyl gauge $a_0 = 0$ to keep only physical degrees of freedom and let us require that the gauge fields become pure gauge configurations $a_i = \partial_i \chi$ on the boundary. Then, we have to promote χ to a full fledged edge degree of freedom governed by a first-order Lagrangian,

$$\mathcal{L}_{\text{edge}} = -\frac{k}{4\pi}\partial_s\chi\partial_t\chi - \mathcal{H} = -\frac{k}{4\pi}\partial_s\chi\left(\partial_t + v\partial_s\right)\chi, \tag{6.27}$$

where s denotes the spatial coordinate along the egde and the Hamiltonian $\mathcal{H} = (k/4\pi)v(\partial_s\chi)^2$, with v the velocity of propagation of the edge mode, describes the non-universal edge mode dynamics and has to be added in by hand. This Lagrangian describes a chiral boson [81], equivalent to a Weyl fermion in (1+1) dimensions. Incompressible quantum fluids have thus massless chiral boson excitations on their edges. Of course, the edge theory of the hierarchical incompressible fluids (6.25) involves N coupled chiral bosons [82].

Chapter 7

The superconductor-to-insulator transition

Electric forces are typically much stronger than magnetic ones. Superconductivity (for a comprehensive review see [5]) is made possible only by getting rid of the Coulomb interaction amongst charge carriers by screening it. As a consequence, the elementary bulk excitations (for type II superconductors, those of interest here) are not anymore the electric charge carriers but rather the magnetic vortices, with an energy per unit length $O(1/\lambda_L^2)$, with λ_L the London penetration depth.

Suppose now that, in a Gedankenexperiment, we "squeeze" a chunk of superconductor along the z direction to become a film of thickness d (for simplicity we shall assume infinite dimensions in the (x, y) plane). What happens in this case to the Coulomb interaction? This is actually a well-known problem in field theory. Formally, it is equivalent to computing the finite-temperature Coulomb potential of instantons in a $(2+1)$-dimensional Abelian gauge theory (see, e.g. [83]). In field theory, a finite temperature $T = 1/\beta$ is introduced by formulating the action on a $(D + 1)$-dimensional Euclidean space-time with the time direction compactified on a circle of length β, with all bosonic fields having periodic boundary conditions (for a review see [59, 84]). At zero temperature, the action for a system of point excitations coupled to gauge fields is governed by the Coulomb potential in $(D + 1)$-dimensional Euclidean space-time, as we have seen in Chapter 5. At finite temperatures, the Coulomb potential is defined, instead, by the periodic delta function

$$\nabla^2 G(|\mathbf{x}|, x^0, \beta) = e^2 \delta^D(\mathbf{x}) \sum_{n=-\infty}^{+\infty} \delta(x^0 - n\beta), \qquad (7.1)$$

with solution

$$G(|\mathbf{x}|, x^0, \beta) = \frac{e^2}{4\pi} \sum_{n=-\infty}^{+\infty} \frac{1}{\sqrt{|\mathbf{x}|^2 + (x^0 - n\beta)^2}}, \tag{7.2}$$

in $D = 2$. The integers n go under the name of Matsubara frequencies [59, 84]. By simple manipulations, we can rewrite this as

$$G(|\mathbf{x}|, x^0, \beta) = \frac{e^2}{4\pi\beta} \sum_{n=-\infty}^{+\infty} \int_0^\infty dt \frac{1}{\sqrt{1+t^2}} \, e^{i\frac{2\pi n}{\beta}\left(t|\mathbf{x}|-x^0\right)}. \tag{7.3}$$

In the regime $x^0 \ll |\mathbf{x}| \ll \beta$ we can keep only the first Matsubara frequencies $n = \pm 1$ and neglect x^0 in the ensuing cosine function, thereby obtaining

$$G(|\mathbf{x}|, x^0, \beta) = G(|\mathbf{x}|, \beta) = \frac{e^2}{2\pi\beta} \int_0^\infty dt \frac{1}{\sqrt{1+t^2}} \, \cos\frac{2\pi}{\beta} t|\mathbf{x}|$$

$$= \frac{e^2}{2\pi\beta} K_0\left(2\pi \frac{|\mathbf{x}|}{\beta}\right), \tag{7.4}$$

where K_0 is the zeroth-order Bessel function of the second kind (MacDonald function) and we have used one if its integral representations.

The same procedure can be applied verbatim to compute the Coulomb potential of charges squeezed into a film of thickness d with periodic boundary conditions. In this case, the periodic delta function represents an infinite series of couples of mirror charges: the effect of these mirror charges is equivalent to imposing periodic boundary conditions. From (7.4), unfortunately, we see that the "squeezed" Coulomb potential becomes singular in the limit $d \to 0$. Let us, however, consider how this "squeezed" potential is modified for a medium with a dielectric permittivity ε. To this end we recall that the potential for a dipole in a dielectric with (relative) permittivity ε depends on the combination ℓ/ε, where ℓ is the dipole length. If we want to maintain the same effect of the infinite couples of mirror charges in a film of fixed thickness d but with a dielectric permittivity ε we must place them at an effective distance εd. Repeating the above calculation, we find that, for $x^0 \ll |\mathbf{x}| \ll \varepsilon d$, the "squeezed" Coulomb potential becomes

$$G_\varepsilon(|\mathbf{x}|, d) = \frac{e^2}{2\pi\varepsilon d} K_0\left(2\pi \frac{|\mathbf{x}|}{\varepsilon d}\right) \approx -\frac{e^2}{2\pi\varepsilon d} \ln\left(\frac{|\mathbf{x}|}{\varepsilon d}\right) + \text{const.} \tag{7.5}$$

This result shows that the singular behaviour is avoided in materials with a very high dielectric permittivity, for which the Coulomb potential in the film becomes genuinely two-dimensional up to a screening length $\lambda_C = \varepsilon d$. An analogous conclusion that a large dielectric permittivity ε increases the range of the two-dimensional Coulomb interaction in a film can be reached for the configuration in which this is sandwiched between two media with smaller permittivities ε_1 and ε_2 [85, 86].

In 2D, the Coulomb interaction is much stronger than in 3D. In particular it has a logarithmic behaviour, like the magnetic interaction $(\pi d/e^2\lambda_L^2)K_0(|\mathbf{x}|/\lambda_L)$ between vortices [5]. But the strength of the Coulomb interaction increases when the film thickness decreases, while the strength of the magnetic vortex interaction decreases. For sufficiently thin films, the Coulomb interaction will again dominate the magnetic interaction and superconductivity will be destroyed in favour of insulating behaviour. Hence the name superconductor-to-insulator (SIT) transition. As always when two orders compete, we witness the formation of textures and structures [87]. This self-organized granularity in the critical vicinity of the quantum transition, first conjectured in [11, 35], has become paradigmatic of the SIT and is supported by scanning tunnelling microscope measurements [88, 89], see Fig. 7.1. Since the SIT is a quantum phase transition,

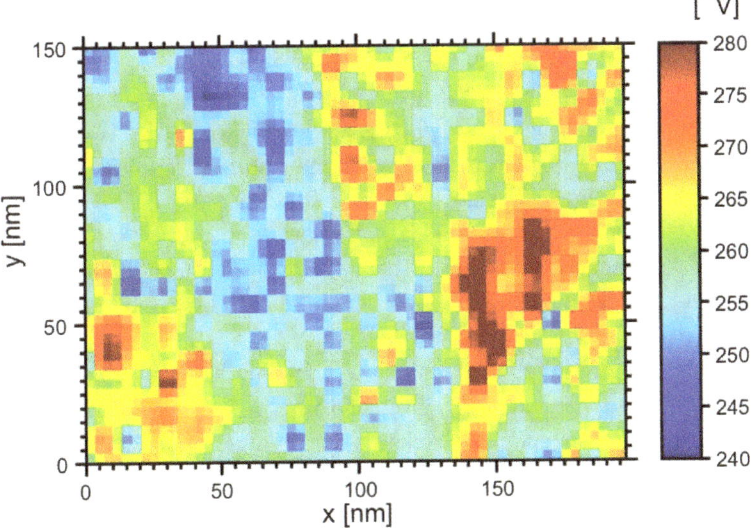

Fig. 7.1. Spatial fluctuations of the order parameter Δ in a TiN film near the SIT, showing structure on the nm scale. From B. Sacépé *et al.*, *Phys. Rev. Lett.* **101**, 157006 (2008), with permission by the authors and the publisher. ©American Physical Society (2008).

which takes places at zero (or extremely low) temperatures as a function of the thickness, the charge carriers remain paired. Since they are localized on nm-scale granules, they can be treated as bosons [9]. The superinsulators and Bose metals proposed in [6] are the two possible quantum states of these out-of-condensate charged bosons.

The SIT can be driven typically both by the thickness of the film or by a magnetic field [9]. A good proxy for the driving parameter is the normal state sheet resistance (resistance per unit area) at a given, fixed temperature [90]. This can be understood as follows. As we have seen, the screening length of the squeezed Coulomb interaction depends only on the scaling variable $d\varepsilon$, with d the thickness and ε the dielectric permittivity of the normal state. Sending $d \to 0$ at fixed screening length requires also $\varepsilon \to \infty$, which, in turns, modifies the normal state resistance. Keeping the screening length fixed, the SIT is thus driven by the simultaneous limits of zero thickness and infinite dielectric permittivity. The SIT has been observed in a wide variety of materials, the first measurements reporting only a direct transition between superconducting and insulating behaviour, identified by a decreasing, respectively increasing sheet resistance when the temperature is lowered. The separatrix of these two behaviours shows metallic behaviour, converging for $T \to 0$ to the universal quantum resistance $h/4e^2 = 6.4$ $k\Omega$, independently of material [9], see Fig. 7.2.

When plotted in terms of the scaling variable $|g - g_c|/f(T)$, with g the driving parameter and g_c its value at the transition, the two branches of the sheet resistance plots collapse on two single curves for the two phases [9], reflecting universality in the vicinity of the quantum transition [68]. The two scaling functions considered most frequently are $f(T) = T^{1/\gamma}$ with $\gamma = 4/3$, corresponding to classical percolation theory, or $\gamma = 7/3$, corresponding to quantum percolation theory. However, we shall see below that experiments actually support a logarithmic scaling function, corresponding formally to $\gamma \to \infty$.

In [6] it was predicted that, under certain conditions, the metallic separatrix between superconductor and insulator could open up to a full fledged metallic phase and that, at sufficiently low temperatures, the insulator is actually a *superinsulator*, with confined electric charge and infinite resistance. This metal made out of bosons, Fig. 7.3, was subsequently called a *Bose metal* [25, 26] and has been detected in many experiments on different materials [28–30] (for a review see [27]). Its very existence poses a fundamental challenge, since orthodoxy has it that metals cannot exist in 2D because of strong Anderson localization [31, 32].

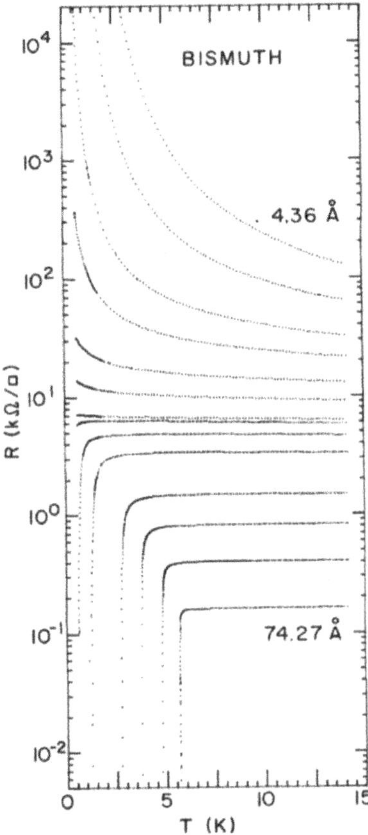

Fig. 7.2. The SIT in a Bi film, as a function of thickness. The separatrix between super-conductor behaviour, with sheet resistance decreasing for decreasing temperature, and insulating behaviour, characterized instead by an increasing sheet resistance, approaches the resistance quantum $h/4e^2 = 6.4\ k\Omega$ for $T \to 0$. From D. Haviland, Y. Liu and A. Goldman, *Phys. Rev. Lett.* **62**, 2180–2183 (1989), with permission by the authors and the publisher. ©American Physical Society (1989).

Superinsulating behaviour has been confirmed by numerical compu-tations [10], was reinvented in [11, 12] and finally detected experimen-tally [11,13,15,22]. It is important to stress that superinsulating behaviour is characterized by an infinite resistance at *finite* temperatures. The usual Arrhenius behaviour

$$R_\square \propto e^{\frac{E_A}{T}}, \tag{7.6}$$

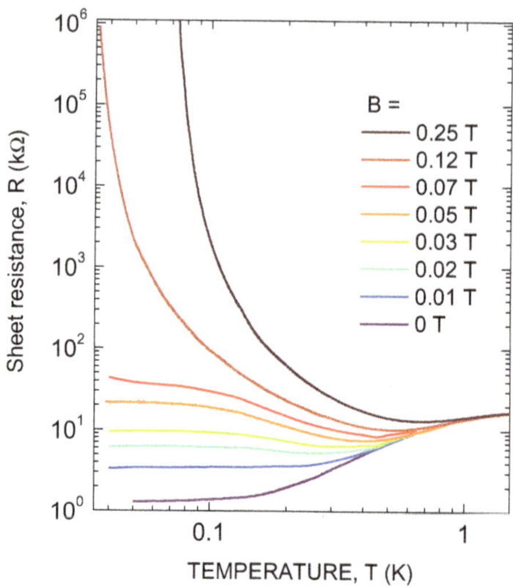

Fig. 7.3. The Bose metal in a NbTiN film, with the SIT driven by a magnetic field. For an entire range of the driving parameter the sheet resistance approaches a constant distributed around the quantum resistance when $T \to 0$, compare with Fig. 7.2. From M. C. Diamantini, A. Yu. Mironov, S. V. Postolova, X. Liu, Z. Hao, D. M. Silevitch, Ya. Kopelevich, P. Kim, C. A. Trugenberger and V. M. Vinokur, Bosonic topological intermediate state in the superconductor–insulator transition, *Phys. Lett.* **A384**, 126570 (2020), ©Elsevier (2020).

with activation energy E_A, characterizing standard insulators, implies that the sheet resistance diverges at $T = 0$. Superinsulators, instead, have a sheet resistance that fits the BKT formula

$$R_\square \propto e^{\sqrt{\frac{aT_{\mathrm{cr}}}{T-T_{\mathrm{cr}}}}}, \tag{7.7}$$

with a a numerical constant. When lowering the temperature, the sheet resistance shows an upturn to hyperactivated behaviour, with $R_\square$ diverging at T_{cr}, see Fig. 7.4, for which $T_{\mathrm{cr}}(B = 0) = 0.062$ mK and $T_{\mathrm{cr}}(B = 0.3 \text{ T}) = 0.175$ mK.

To conclude this chapter we would like to mention that the SIT is often considered a quantum transition driven by structural disorder in the superconducting materials. This idea has its origin in [9], in which it was posited that structural disorder can localize bosons in 2D. This localization manifests as an inhomogenous superconducting condensate which decomposes into superconducting granules [35,88,89] coupled by Josephson links, a picture that was confirmed by numerical simulations [91]. Superconductivity

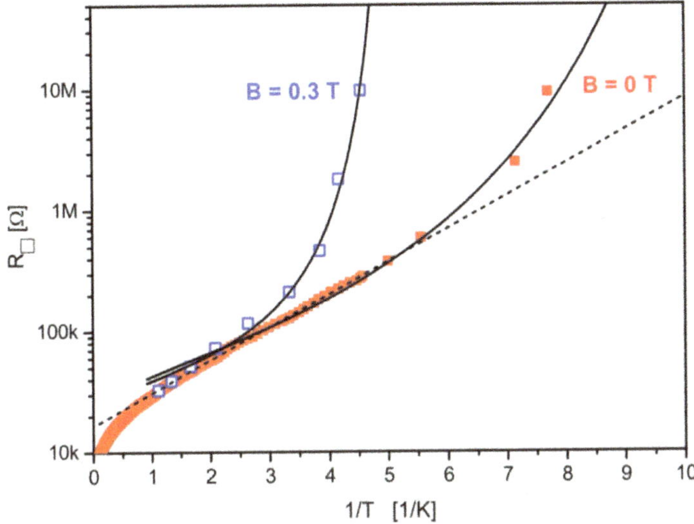

Fig. 7.4. Sheet resistance as a function of inverse temperature for a TiN film. The dashed straight line in this logarithmic scale is the Arrhenius behaviour typical of insulators. The deviations at very low temperatures show the hyperactivated behaviour characterizing superinsulators with $T_{\mathrm{cr}}(B = 0) = 0.062$ mK and $T_{\mathrm{cr}}(B = 0.3\ \mathrm{T}) = 0.175$ mK. From T. I. Baturina and V. M. Vinokur, *Ann. Phys.* **331**, 236–257 (2013), with permission by the authors and the publisher, ©Elsevier (2013).

breaks down entirely in favour of insulating behaviour when these links are lost.

Another role of disorder which is often posited as the origin of the all-important granular structure is Anderson localization (for a review see [92]). This conception is inaccurate. To illustrate this, let us first consider a very similar system where it is also believed that structural disorder plays a paramount role, the quantum Hall effect. In the quantum Hall effect the role of impurities is to break translation invariance, so that the Landau levels are broadened and the observed plateaus form [93]. It is generally thought that, without structural disorder, only the classical Hall effect would be possible. However, while structural disorder may help to broaden the plateaus, the ground states that form at the magical filling fractions corresponding to these plateaus are topological and thus completely transparent to disorder. They behave as if disorder would not exist. Even more importantly, the notion that structural disorder is necessary for plateaus has been recently dispelled [94]. The necessary translation invariance breaking is provided by the formation of a Wigner crystal in the vicinity of the magic filling fractions and the quantum Hall effect can take place in complete absence of disorder. Structural disorder is thus reduced to a supporting role, it can help plateau formation but is not necessary for this.

Something similar happens in the SIT. The SIT point is universal, as is known since the work of Fisher [9] and the resistance quantum at this point is a longitudinal equivalent to the transverse resistance quantum of the integer quantum Hall effect. As we show below, the universal ground state at this point can also broaden to a full fledged intermediate topological state, the equivalent of a plateau, due to purely topological interactions among charges and vortices. As we have shown above, in 2D charges and vortices have the same logarithmic interactions. When their strength becomes comparable by decreasing the film thickness, there is a strong competition that prevents both a charge and a vortex condensate. In this case, the infrared dominant interaction between charges and vortices, the one that determines the character of the ground state, is the topological mutual statistics interaction encoding the Aharonov–Bohm and Aharonov–Casher phases: the ground state in this intermediate "plateau" is, correspondingly, a topological ground state, in which disorder and all types of localization become irrelevant. Superconductivity is destroyed in a quantum Berezinskii–Kosterlitz–Thouless transition by the competition between two quantum orders, a charge order and a vortex order. This competition gives rise to an emergent granular structure, with no need whatsoever of structural disorder. As in the above example of the quantum Hall effect, structural disorder can help in forming the localized granular structure and tune the strength of the Coulomb interaction [95] but is in no way instrumental for it. The SIT is not driven by disorder but mainly by the topology-driven competition between two distinct quantum orders.

Finally, let us remark that Anderson localization does not survive linear interactions as those in the superinsulating state, not even in 1D, where they are kinematic, let alone in 2D. While disorder may play a role in tuning the strength of the Coulomb interaction, localization is not relevant for the nature of the phases around the SIT.

Chapter 8

Effective field theory of the SIT tricritical point

As we have discussed in Chapter 7, the degrees of freedom in the vicinity of the SIT are Cooper pairs localized on granules of minimum size ℓ and Josephson-like vortices in between. This information is sufficient to establish the entire phase diagram. Charges and vortices acquire Aharonov–Bohm [44] and Aharonov–Casher [54] phases when one type of excitation encircles the other. These mutual statistics interactions (for a review see [79]), moreover, are the infrared (IR)-dominant ones since they have infinite range. In the Euclidean partition function describing point charges and vortices they must be accounted for by the topological Gauss linking number of their trajectories (3.27), with the value $k = 1$ to guarantee the Dirac quantization condition [17]. Unfortunately, this topological invariant is non-local. It was, however, noted by Wilczek [55] that it can be made local by coupling the point charge and vortex currents Q_μ and M_μ to two fictitious gauge fields a_μ and b_μ with mixed Chern–Simons term, giving rise to the Euclidean partition function

$$Z = \sum_{\{Q_\mu, M_\mu\}} \int \mathcal{D}a_\mu \mathcal{D}b_\mu \; e^{-S},$$

$$S = \int d^3x \left(\frac{i}{2\pi} a_\mu \epsilon^{\mu\alpha\nu} \partial_\alpha b_\nu + i a_\mu Q_\mu + i b_\mu M_\mu \right). \tag{8.1}$$

This same model has also another, albeit equivalent interpretation. Following Wen and Zee in their seminal derivation of the long-distance effective field theories of incompressible quantum Hall states [76], as discussed in Chapter 6, let us define topologically conserved charge and vortex currents in terms of the curl of a fictitious pseudovector field b_μ and a fictitious vector field a_μ,

$$j^\mu = \frac{1}{2\pi}\epsilon^{\mu\alpha\nu}\partial_\alpha b_\nu,$$

$$\phi^\mu = \frac{1}{2\pi}\epsilon^{\mu\alpha\nu}\partial_\alpha a_\nu. \tag{8.2}$$

The fields a_μ and b_μ are gauge fields, since physical currents are left invariant by the gauge transformations $a_\mu \to a_\mu + \partial_\mu\xi$ and $b_\mu \to b_\mu + \partial_\mu\lambda$. The effective action for states of matter involving charges and vortices must therefore be a gauge theory. The IR-dominant gauge invariant term which preserves the parity ($\mathcal{P}$) and time-reversal ($\mathcal{T}$) symmetries is the mixed Chern–Simons term. If the gauge group is U(1), the gauge fields are compact, with gauge functions being angular variables defined in the domain $[-\pi, +\pi]$. Remembering the identity (2.15), we recognize that, exactly as in the case of the XY model, a compact mixed Chern–Simons term involves both a fluctuation term and a topological excitation term. The topological excitations are singular "electric" and "magnetic" strings Q_μ and M_μ representing the (Euclidean) world-lines of charges and vortices. When these are taken into account separately, we get back (8.1). As a consequence, this model can also be interpreted simply as a compact mixed Chern–Simons model.

The Chern–Simons term encodes only the topological interactions. To include dynamics we will add all possible local, gauge invariant terms. In addition, to take into account the discrete nature of the granules, we will formulate the model on a lattice of spacing ℓ. This leads us to the most general Euclidean effective action for the vicinity of the SIT [6],

$$S = \sum_x i\frac{\ell_0\ell^2}{2\pi}a_\mu k_{\mu\nu}b_\nu + \frac{\ell_0\ell^2}{2e_v^2\mu}f_0^2 + \frac{\ell_0\ell^2\varepsilon}{2e_v^2}\mathbf{f}^2 + \frac{\ell_0\ell^2}{2e_q^2\mu}g_0^2 + \frac{\ell_0\ell^2\varepsilon}{2e_q^2}\mathbf{g}^2$$

$$+ \sum_{x,i} i\ell_0 a_0 Q_0 + i\ell a_i Q_i + i\ell_0 b_0 M_0 + i\ell b_i M_i, \tag{8.3}$$

where ℓ_0 is the lattice spacing in the Euclidean time direction, $k_{\mu\nu}$ is the lattice Chern–Simons term introduced in Chapter 4 and the fluctuation currents become

$$j^\mu = \frac{1}{2\pi}f^\mu = \frac{1}{2\pi}k_{\mu\nu}b_\nu,$$

$$\phi^\mu = \frac{1}{2\pi}g^\mu = \frac{1}{2\pi}k_{\mu\nu}a_\nu. \tag{8.4}$$

The integer link variables Q_μ and M_μ represent point charges and vortices as described above.

There are four phenomenological parameters in this action. The couplings e_v^2 and e_q^2 distinguish between the electric and magnetic sectors of the theory and have canonical dimension [1/length]. The dimensionless parameters ε and μ distinguish between "electric" and "magnetic" components within a sector and represent the relative dielectric permittivity and magnetic permeability, as we will show below. Correspondingly, $v = 1/\sqrt{\mu\varepsilon}$ is the velocity of light in the material. As discussed in Chapter 7, we are interested in materials in which $v \ll 1$ because "magnetic" components are suppressed with respect to "electric" ones. This is realized in the two extreme cases $\mu = 1$, $\varepsilon \gg 1$ or $\varepsilon = 1$, $\mu \gg 1$. The former case corresponds to superconducting thin films, the latter one to the fabricated metamaterials to be presented in a later chapter. In the rest of this chapter we shall thus set $\mu = 1$ and consider large values of ε. By multiplying the whole action by $1/\sqrt{\varepsilon}$ we obtain the discrete version of the non-relativistic doubled topologically massive gauge theory (3.28) discussed in Chapter 3,

$$S = \sum_x i\frac{\ell_0\ell^2}{2\pi}a_\mu k_{\mu\nu}b_\nu + \frac{\ell_0\ell^2 v}{2e_v^2}f_0^2 + \frac{\ell_0\ell^2}{2e_v^2 v}\mathbf{f}^2 + \frac{\ell_0\ell^2 v}{2e_q^2}g_0^2 + \frac{\ell_0\ell^2}{2e_q^2 v}\mathbf{g}^2$$

$$+ \sum_{x,i} i\ell_0 a_0 Q_0 + i\ell a_i Q_i + i\ell_0 b_0 M_0 + i\ell b_i M_i. \tag{8.5}$$

Finally, the same rescalings leading to (3.29) lead also here to the simple form

$$S = \sum_x i\frac{\ell^3}{2\pi}a_\mu k_{\mu\nu}b_\nu + \frac{\ell^3}{2e_v^2}f_\mu f_\mu + \frac{\ell^3}{2e_q^2}g_\mu g_\mu + i\ell a_\mu Q_\mu + i\ell b_\mu M_\mu, \tag{8.6}$$

where we have chosen the appropriate value of the Euclidean lattice spacing as $\ell_0 = \ell/v$. In this formulation we are left only with the two phenomenological parameters distinguishing between the electric and magnetic sectors of the model. Recalling that the dual field strengths of the two gauge fields encode the electric and magnetic currents, we recognize that these parameters represent the electric and magnetic energies of a fundamental granule, as discussed in Chapter 7,

$$e_q^2 = O\left(\frac{e^2}{2\pi d}\right),$$

$$e_v^2 = O\left(\frac{\pi}{e^2\lambda_\perp}\right) = O\left(\frac{\pi d}{e^2\lambda_L^2}\right), \tag{8.7}$$

where $\lambda_\perp = \lambda_L^2/d$ is the Pearl length [5]. If the effects of the topological excitations are neglected, both charges and vortices acquire a dispersion relation

$$E = \sqrt{m^2 v^4 + v^2 p^2}, \tag{8.8}$$

with topological mass

$$m = \frac{e_q e_v}{2\pi v} = O\left(\frac{1}{v\lambda_L}\right). \tag{8.9}$$

Given that e_q^2 and e_v^2 have canonical dimension [1/length] and, correspondingly, the two kinetic terms in (8.6) are infrared-irrelevant, one might be tempted to leave them out and consider anyway only the infrared-dominant mixed Chern–Simons theory. Unfortunately, this leads to wrong results and the origin of the problem lies exactly in the role of the topological excitations due to the compactness of the two gauge groups. The two parameters e_q^2 and e_v^2 can be traded for the topological mass $m = e_q e_v / 2\pi v$ and one dimensionless parameter $g = e_v/e_q = O(d/(\alpha\lambda_L))$, where $\alpha = e^2/4\pi$ is the fine structure constant. As we will show later, this parameter plays the role of a dimensionless conductivity. For the moment, the important point is that the "topological limit" $e_q^2 \to \infty$ and $e_v^2 \to \infty$ is not well defined without specifying the value of g in this limit. As we now show, the behaviour of the topological excitations depends crucially on this value and, as a consequence, one obtains very different ground states when g is varied.

Let us integrate out the fictitious gauge fields to obtain an action for the topological excitations alone. To this end we use the gauge fixing technique explained in Chapter 3, obtaining

$$
\begin{aligned}
S_{\text{top}} = \sum_x v^2 \frac{e_q^2}{\ell} Q_\mu & \frac{1}{v^4 m^2 - d_0 \hat{d}_0 - v^2 \nabla_2^2} Q_\mu \\
+ v^2 \frac{e_v^2}{\ell} M_\mu & \frac{1}{v^4 m^2 - d_0 \hat{d}_0 - v^2 \nabla_2^2} M_\mu \\
+ i \frac{2\pi v^6 m^2}{\ell} Q_\mu & \frac{k_{\mu\nu}}{(d_0 \hat{d}_0 + v^2 \nabla_2^2)(v^4 m^2 - d_0 \hat{d}_0 - v^2 \nabla_2^2)} M_\mu,
\end{aligned}
\tag{8.10}
$$

where ∇_2 is the 2D spatial Laplacian. The third term in this action describes the lattice version of the topological linking of electric and magnetic strings of width $1/(vm)^2$, as in (3.24). Because of the Dirac quantization condition, at large distances, i.e. in the limit $vm \to \infty$, it reduces to an integer. This can be easily recognized by representing the transverse electric strings as $Q_\mu = k_{\mu\nu} S_\nu$ with $d_\nu S_\nu = 0$, this last gauge condition ensuring that S_μ contains the same number of degrees of freedom as the original Q_μ. Summing by parts and using (4.5) one can rewrite the entire linking term as $i2\pi \times$ integer. We shall thus henceforth drop this term and focus only on the electric–electric and magnetic–magnetic terms in (8.10).

In order to proceed we follow the standard lattice gauge field theory arguments of [96] to retain only the self-interaction terms in (8.10). Consider strings made of N bonds with integer electric and magnetic quantum numbers Q and M. Such configurations can be assigned an action

$$S_{\text{top}} = 2\pi m\ell v G(m\ell v) \left[\frac{e_q}{e_v} Q^2 + \frac{e_v}{e_q} M^2 \right] N, \tag{8.11}$$

where $G(m\ell v)$ is proportional to the diagonal element of the lattice kernel $G(m\ell v, x-y)$ representing the inverse of the operator $(\ell^2/v^2)(m^2 v^4 - d_0 \hat{d}_0 - v^2 \nabla_2^2)$. We shall take $G(m\ell v) = rG(m\ell v, x-y)$, where the numerical factor r encodes the corrections to the self-interaction approximation. The kernel $G(m\ell v, x)$ is defined by the equation

$$\frac{\ell^2}{v^2}(m^2 v^4 - d_0 \hat{d}_0 - v^2 \nabla_2^2)\, G(m\ell v, x) = \delta_{x,0}. \tag{8.12}$$

Using the Fourier transform

$$G(m\ell v, x) = \frac{\ell^3}{v} \int_{-\pi v/\ell}^{+\pi v/\ell} dk_0 \int_{-\pi/\ell}^{+\pi/\ell} d^2 k\, G(m\ell v, k) \exp(ik \cdot x), \tag{8.13}$$

we obtain

$$\frac{\ell^3}{v} \int_{-\pi v/\ell}^{+\pi v/\ell} dk_0 \int_{-\pi/\ell}^{+\pi/\ell} d^2 k\, G(m\ell v, k) \frac{\ell^2}{v^2}(m^2 v^4 - d_0 \hat{d}_0 - v^2 \nabla_2^2) e^{ik \cdot x}$$

$$= \frac{1}{(2\pi)^3} \int_{-\pi}^{\pi} d^3 k\, e^{ik \cdot x}. \tag{8.14}$$

Applying finally the operator $(\ell^2/v^2)(m^2 v^4 - d_0 \hat{d}_0 - v^2 \nabla_2^2)$ to the exponential in the Fourier transform and rescaling momenta gives the final result

$$G(m\ell v) = \frac{r}{(2\pi)^3} \int_{-\pi}^{\pi} d^3 k \frac{1}{(m\ell v)^2 + \sum_{i=0}^{2} 4 \sin\left(\frac{k^i}{2}\right)^2}. \tag{8.15}$$

The action (8.11) in a Euclidean field theory model plays the same role as the energy in an equivalent statistical mechanics model in one additional spatial dimension. Quantum corrections to the classical action play the same role as the entropy in the equivalent statistical mechanics model. This string "entropy" is also proportional to their length, being given by μN with $\mu \approx \ln(5)$ since, at each step, the non-backtracking strings can choose among 5 possible directions how to continue. One can thus assign the "free energy"

$$F = 2\pi m\ell v G(m\ell v) \left[\frac{e_q}{e_v} Q^2 + \frac{e_v}{e_q} M^2 - \frac{1}{\eta} \right] N \qquad (8.16)$$

to a string of length $L = \ell N$ carrying electric and magnetic quantum numbers Q and M, respectively. Here we have introduced the dimensionless parameter

$$\eta = \frac{2\pi m\ell v G(m\ell v)}{\mu}, \qquad (8.17)$$

which, together with the ratio $g = e_v/e_q$, fully determines the quantum phase structure.

The ground state of the quantum model is found by minimizing its free energy as a function of N. When the energy term in (8.16) dominates, the free energy is positive and consequently minimized by short closed loop configurations. When, instead, the entropy dominates, the free energy is negative and minimized by infinitely long strings and large loops. The condition for condensation of long strings with integer quantum numbers Q and M is thus given by

$$\eta \frac{e_q}{e_v} Q^2 + \eta \frac{e_v}{e_q} M^2 < 1. \qquad (8.18)$$

If two or more condensations are allowed, one has to choose the one with the lowest free energy. This condition describes the interior of an ellipse with semiaxes

$$r_Q = \sqrt{\frac{e_v}{e_q}} \sqrt{\frac{1}{\eta}},$$
$$r_M = \sqrt{\frac{e_q}{e_v}} \sqrt{\frac{1}{\eta}}, \qquad (8.19)$$

on a square lattice of integer electric and magnetic charges. The ratio $g = e_v/e_q$ determines the ratio of the semiaxes, the parameter η sets the overall scale of the ellipse. The quantum phase diagram is found by noting which integer charges lie within the ellipse when the semiaxes and the overall scale are varied, as shown in Fig. 8.1,

$$\eta < 1 \rightarrow \begin{cases} g > 1, \text{ electric condensation } = \text{ superconductor,} \\ g < 1, \text{ magnetic condensation } = \text{ superinsulator,} \end{cases}$$

$$\eta > 1 \rightarrow$$

$$\begin{cases} g > \eta, \text{ electric condensation } = \text{ superconductor,} \\ \eta > g > \frac{1}{\eta}, \text{ no condensation } = \text{ Bose metal } = \text{ topological insulator,} \\ g < \frac{1}{\eta}, \text{ magnetic condensation } = \text{ superinsulator,} \end{cases}$$

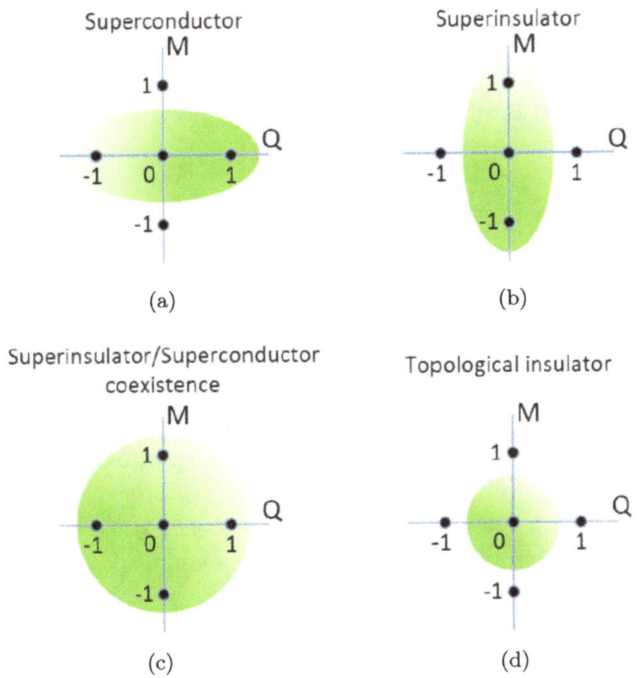

Fig. 8.1. The ellipse in parameter space determining the quantum phase diagram of the SIT. From M. C. Diamantini, A. Yu. Mironov, S. V. Postolova, X. Liu, Z. Hao, D. M. Silevitch, Ya. Kopelevich, P. Kim, C. A. Trugenberger and V. M. Vinokur, Bosonic topological intermediate state in the superconductor–insulator transition. *Phys. Lett.* **A384**, 126570 (2020), ©Elsevier (2020).

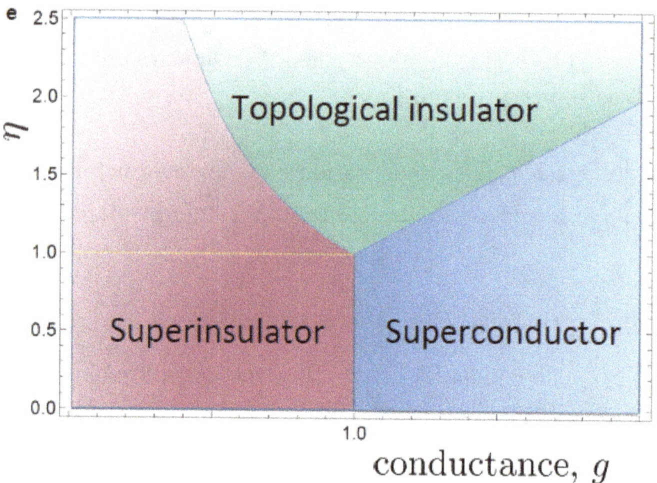

Fig. 8.2. The quantum phase diagram of the SIT, with the tricritical point at $g = 1$, $\eta = 1$ dominating the phase structure. From M. C. Diamantini, A. Yu. Mironov, S. V. Postolova, X. Liu, Z. Hao, D. M. Silevitch, Ya. Kopelevich, P. Kim, C. A. Trugenberger and V. M. Vinokur, Bosonic topological intermediate state in the superconductor–insulator transition. *Phys. Lett.* **A384**, 126570 (2020), ©Elsevier (2020).

where we have anticipated that an electric condensation phase is a superconductor, the dual magnetic condensation phase is a superinsulator and the phase with no condensations is a Bose metal, a.k.a bosonic topological insulator [97], as we will show later.

As anticipated, the quantum phase diagram, shown in Fig. 8.2, is fully determined by the dimensionless conductance (see below) $g = O(d/\alpha\lambda_L)$ and by the parameter $mv\ell = O(\ell/\lambda_L)$, where $\alpha = e^2/4\pi$ is the fine structure constant. The SIT is driven by the thickness d of the film. For thin films with $d \approx \xi$, with ξ the coherence length, the quantum coupling $g = O(1/\kappa\alpha)$, with $\kappa = \lambda_L/\xi$ the Ginzburg–Landau parameter of the superconductor. Strongly type II superconductors require thus less thin films to observe the SIT.

The parameter η determines if the SIT is a direct first-order transition (with a coexistence regime) between superconductor and superinsulator or if it goes via an intermediate phase, which, as we will discuss in detail below, constitutes a bosonic topological insulator [97]. This new parameter, shown in Fig. 8.3, is a function of the ratio of the size ℓ of the fundamental granule to the London penetration length. Only materials with these two sizes of comparable magnitude can show an intermediate phase.

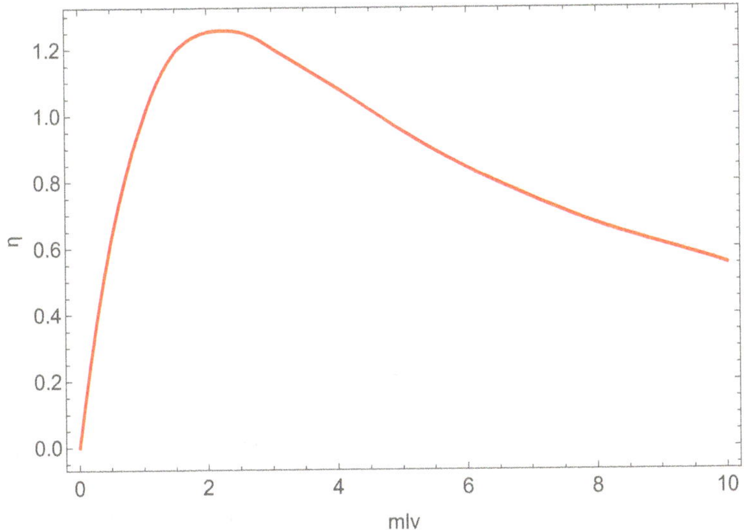

Fig. 8.3. The parameter $\eta(mv\ell)$ for the example $r = 1.5$.

While the direct SIT is a first-order transition, going through a phase-separated coexistence regime, we will show below that the two transitions between superinsulator and topological insulator and between topological insulator and superconductor at $\eta > 1$ are continuous transitions which become BKT transitions [14] in the deep non-relativistic limit. The point $g = 1$, $\eta = 1$ is thus a quantum tricritical point. In the vicinity of this tricritical point the two dynamical scales d and ℓ determining the phase diagram satisfy

$$\ell \approx \frac{d}{\alpha}, \qquad (8.20)$$

which shows that the formation of granules when squeezing the material into a film is due to strong quantum fluctuations in this limit.

Before we delve into the derivation of the detailed properties of the various phases near the SIT tricritical point, let us spend some words on the magnetic-field-driven SIT and on the finite-temperature phase transitions. In concrete experimental settings the phase transitions in thin films are driven either by varying the thickness of the films or by varying an applied magnetic field. In this situation there is one additional parity ($\mathcal{P}$) and time reversal ($\mathcal{T}$) breaking external parameter that must be accounted for in the gauge theory formulation.

A uniform external magnetic field can be simply incorporated by the minimal coupling $i\ell^3 (2e/2\pi) A_\mu f_\mu$ to the charge current $j_\mu = (1/2\pi) f_\mu$, Eq. (8.2). By a summation by parts, this amounts to the following additional term in the gauge theory action (8.6),

$$S \to S + \sum_x i\ell \, b_0 f, \tag{8.21}$$

where f denotes the number of elementary fluxes π/e per plaquette piercing the lattice. This modification amounts simply to shifting the integers

$$M_\mu \to M_\mu + \Phi_\mu, \tag{8.22}$$

in the original gauge theory model, where Φ_μ represents infinitely long strings in the Euclidean time direction at each lattice point, such that $\Phi_i = 0$ and $\Phi_0 = f$. This shows that f is a periodic parameter defined modulo an integer: it is called the magnetic frustration, $0 \le f < 1$, which explains why it is usually denoted by f. An external magnetic field corresponds thus to a special case of condensed magnetic strings with non-integer quantum number, giving rise to the modified string free energy

$$F = 2\pi m\ell v G(m\ell v) \left[\frac{e_q}{e_v} \, Q^2 + \frac{e_v}{e_q} \, (M+f)^2 - \frac{1}{\eta} \right] N. \tag{8.23}$$

To compare with the experimentally relevant situation let us assume we start with $f = 0$ in the superconducting phase with condensation of electric strings,

$$\begin{aligned} \eta < 1 \;\; &\to g > 1, \\ \eta > 1 \;\; &\to g > \eta. \end{aligned} \tag{8.24}$$

In this case, the original $f = 0$ ellipse is elongated along the electric quantum number axis. Turning on an external magnetic field amounts to increasing the frustration parameter f and, as consequence, the ellipse moves down along the magnetic quantum number axis, as shown in Fig. 8.4.

There are two important thresholds in this downward movement. The points on the ellipse corresponding to $Q = \pm 1$ have M coordinate $M = r_M \sqrt{1 - (1/r_Q^2)}$, where r_Q and r_M are the semiaxes. The M-negative ellipse point with $Q = 0$, instead, lies at a distance r_M from the origin by definition. A transition can be caused either by the fact that the two points $Q = \pm 1$ will "exit" the interior of the ellipse before the point $M = -1$ has "entered" it, in which case the initial superconductor turns into a bosonic

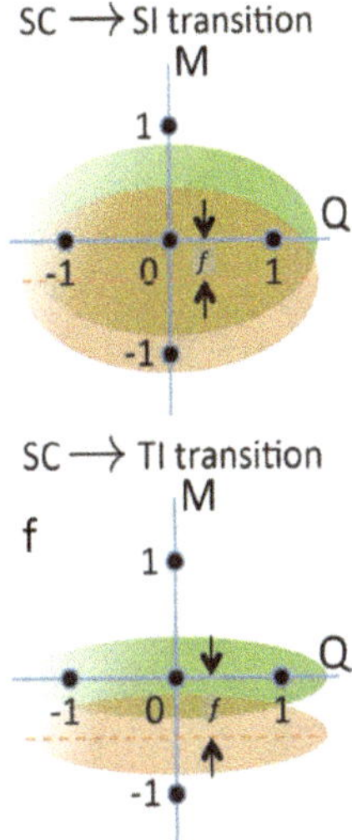

Fig. 8.4. Downward shift of the parameter ellipse by an external magnetic field. From M. C. Diamantini, A. Yu. Mironov, S. V. Postolova, X. Liu, Z. Hao, D. M. Silevitch, Ya. Kopelevich, P. Kim, C. A. Trugenberger and V. M. Vinokur, Bosonic topological intermediate state in the superconductor–insulator transition. *Phys. Lett.* **A384**, 126570 (2020), ©Elsevier (2020).

topological insulator/Bose metal, or by the fact that the point $M = -1$ has "entered" the ellipse interior while the two points $Q = \pm 1$ are still inside. In this case we have a coexistence regime of electric and magnetic strings. There will thus be a first-order direct transition to a superinsulator when the frustration reaches a point such that the energy of the magnetic strings becomes smaller than that of the electric ones. Clearly, the two thresholds are given by the frustrations

$$f_1 = r_M \sqrt{1 - \frac{1}{r_Q^2}},$$

$$f_2 = 1 - r_M,$$

(8.25)

and the condition for an intermediate bosonic topological insulator/Bose metal phase is $f_1 < f_2$. Using the explicit expressions (8.19) for the semiaxes in terms of lattice parameters we can translate this condition into

$$F(g) \equiv \frac{1}{g^2} - \frac{2}{\sqrt{g\eta}} + 1 > 0.$$

(8.26)

Let us first consider the case $\eta < 1$ and let us express

$$\frac{1}{g} = 1 - \epsilon, \tag{8.27}$$

to satisfy (8.24). In this case the function F reduces to

$$F(g) = 2\left(1 - \frac{1}{\sqrt{\eta}}\right) - O(\epsilon) < 0, \tag{8.28}$$

which shows that, in this case, there is a direct transition from a superconductor to a superinsulator. For small f, the transition takes place when the new effective magnetic semiaxis $r_M = r_M(1 + f)$ of the ellipse becomes larger than the originally larger electric semiaxis r_Q. This gives the critical frustration

$$f_{\text{crit}} = \frac{1}{2}\left(g^2 - 1\right). \tag{8.29}$$

Let us now consider the second case $\eta > 1$. In this case (8.24) requires

$$\frac{1}{g} = \frac{1}{\eta} - \epsilon, \tag{8.30}$$

and the function F becomes

$$F(g) = \left(\frac{1}{\eta} - 1\right)^2 - O(\epsilon) > 0, \tag{8.31}$$

confirming that, in this case, instead the transition from the superconductor is to a bosonic topological insulator/Bose metal phase. In this case the critical frustration is determined by the value at which the original electric topological excitations exit the shifted ellipse,

$$f_{\text{crit}} = f_1 = \frac{1}{\sqrt{g}}\sqrt{\frac{1}{\eta} - \frac{1}{g}}. \tag{8.32}$$

To conclude this chapter, let us ask the question of what happens when we raise the temperature. In field theory, a finite temperature T is introduced by formulating the action on a Euclidean time of finite length $\beta = 1/T$, with periodic boundary conditions for bosonic fields, such as in the present case (we reabsorb the Boltzmann constant into the temperature). If the original field theory model is defined on a Euclidean lattice of spacing ℓ, with a propagation velocity v, then β is quantized in integer multiples of $\ell_0 = \ell/v$. We shall define an integer, dimensionless temperature b by $\beta = (2\ell/v)b$. Because of the lattice structure, energies are defined only within a Brillouin zone of length $2\pi v/\ell$ but on the other side, due to the periodic boundary condition in the Euclidean time direction, the energy k^0 must be also quantized in integer multiples of $2\pi/\beta$. This can be achieved simultaneously by considering discrete values $k^0 = (\pi v/\ell b)k$, with the integers $k \in \mathbb{Z}$ known as Matsubara frequencies, and substituting

$$\int_{-\frac{\pi v}{\ell}}^{\frac{\pi v}{\ell}} dk^0 f\left(k^0\right) \rightarrow \sum_{k=-b}^{k=b} \frac{\pi v}{\ell b} f\left(\frac{\pi vk}{b\ell}\right). \tag{8.33}$$

In our concrete case, the finite temperature $T > 0$ affects the scale parameter η of the ellipse via the coefficient $G(m\ell v)$. At zero temperature this is given by

$$G(m\ell v) = \frac{r}{(2\pi)^3} \int_{-\pi}^{\pi} d^3k \frac{1}{(m\ell v)^2 + \sum_{i=0}^{2} 4 \sin\left(\frac{k^i}{2}\right)^2}. \tag{8.34}$$

At finite temperatures it has to be modified according to (8.33),

$$G(m\ell v, T) = \frac{r}{(2\pi)^3} \sum_{k=-b}^{k=+b} \frac{\pi}{b} \int_{-\pi}^{\pi} \frac{dk^1 dk^2}{(m\ell v)^2 + 4 \sin\left(\frac{\pi k}{2b}\right)^2 + \sum_{i=1}^{2} 4 \sin\left(\frac{k^i}{2}\right)^2}. \tag{8.35}$$

By a numerical computation one can verify over 3 orders of magnitude ($m\ell v = 0.001$ to $m\ell v = 1$) covering the relevant range that the ratio $S(T) = G(m\ell v, T)/G(m\ell v)$ does not depend on the parameter $m\ell v$ but is rather a function of the temperature alone. As a consequence, η and the semiaxes of the ellipse scale with the inverse of the function $S(T)$. This means that, with increasing temperature, the whole ellipse shrinks by the scale factor $S(T)$, as shown in Fig. 8.5. Quantum numbers $M = \pm 1$ that are within the ellipse at $T = 0$, will thus "exit" its interior at some critical temperature defined by the condition

$$\frac{1}{g\eta} = S(T_{\text{CBKT}}), \tag{8.36}$$

assuming that the quantity on the left-hand side is larger than one (i.e. that there is a superinsulator at $T = 0$). Since the magnetic semiaxis is always longer and thus no electric quantum numbers may appear within the ellipse interior when the magnetic ones have fallen outside, the superinsulator experiences a direct deconfinement transition to an insulator at $T = T_{\text{CBKT}}$. The subscript indicates that this charge-related transition is in the BKT [14]

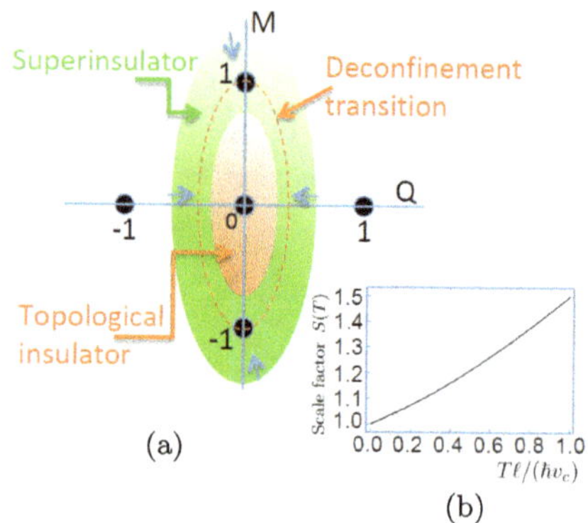

Fig. 8.5. The superinsulator deconfinement transition at a critical temperature T_{CBKT}. Panel (a): shrinking of the ellipse. Panel (b): the scale factor $S(T)$. From M. C. Diamantini, C. A. Trugenberger, V. M. Vinokur, Confinement and asymptotic freedom with Cooper pairs, *Comm. Phys.* **1**(77), 1 (2018). Creative Commons Attribution 4.0.

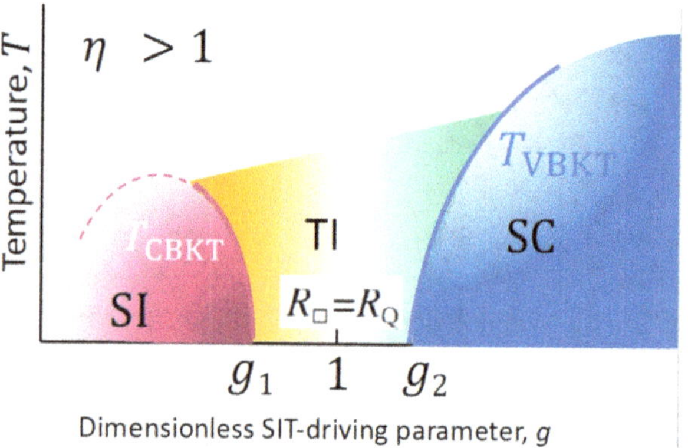

Fig. 8.6. Finite-temperature phase diagram of the SIT for the case $\eta > 1$. The sheet resistance $R_\square$ at the self-dual point corresponds to the resistance quantum R_Q. From M. C. Diamantini, A. Yu. Mironov, S. V. Postolova, X. Liu, Z. Hao, D. M. Silevitch, Ya. Kopelevich, P. Kim, C. A. Trugenberger and V. M. Vinokur, Bosonic topological intermediate state in the superconductor–insulator transition. *Phys. Lett.* **A384**, 126570 (2020), ©Elsevier (2020).

universality class, as we will show below. Correspondingly, superconductors undergo a vortex BKT phase transition at T_{VBKT} defined by

$$\frac{g}{\eta} = S(T_{\mathrm{VBKT}}). \tag{8.37}$$

The resulting finite-temperature phase diagram is shown in Fig. 8.6.

Chapter 9

The nature of the phases in the vicinity of the SIT tricritical point

In what follows we shall compute the electromagnetic response in the various quantum phases of the system at $T = 0$ in order to establish their nature. To accomplish this we minimally couple the current $j_\mu = (1/2\pi)k_{\mu\nu}b_\nu$ to the physical electromagnetic gauge field A_μ in the Euclidean action,

$$S \to S + i \sum_x \ell_0 \ell^2 A_\mu j_\mu = S + i \sum_{x,i} \frac{\ell}{2\pi} \left(b_0 \ell^2 F_0 + b_i \ell_0 \ell F_i \right), \qquad (9.1)$$

where $F_\mu = \hat{k}_{\mu\nu} A_\nu$ is the dual electromagnetic field strength and b_μ is the rescaled gauge field appearing in (8.6), and we compute the electromagnetic response, defined by the effective action

$$\mathrm{e}^{-S_{\mathrm{eff}}(A_\mu)} = \sum_{Q_\mu, M_\mu} \int_{a_\mu, b_\mu} \mathcal{D}a_\mu \mathcal{D}b_\mu \, \mathrm{e}^{-S(a_\mu, b_\mu, Q_\mu, M_\mu, A_\mu)}. \qquad (9.2)$$

The various phases of the system can then be distinguished by this effective action or, equivalently, by the current induced by external electromagnetic fields,

$$j_\mu = \frac{1}{\ell_0 \ell^2} \frac{\delta}{\delta A_\mu} S_{\mathrm{eff}}(A_\mu). \qquad (9.3)$$

Comparing (9.1) with the coupling $\sum_x \ell b_\mu M_\mu$ in (8.6) we see that a physical electromagnetic gauge field can be simply taken into account by a shift $M_0 \to M'_0 = M_0 + (1/2\pi)\ell^2 F_0$, $M_i \to M'_i = M_i + (1/2\pi)\ell_0 \ell F_i$. This gives

$$\mathrm{e}^{-S_{\mathrm{eff}}(A_\mu)} = \sum_{Q_\mu, M_\mu} \mathrm{e}^{-S_{\mathrm{top}}(Q_\mu, M'_\mu)}, \qquad (9.4)$$

73

where S_{top} is the action (8.10). Retaining here too, as in (8.11), only self-interaction terms, this reduces to the local form

$$S_{\text{top}}\left(Q_\mu, M'_{\mu}\right)$$

$$= \sum_{x,i} \ell_0 \ell^2 \frac{e_v}{e_q} \frac{\mu\eta\ell}{4\pi^2} \left[v\left(F_0 + \frac{2\pi}{\ell^2} M_0\right)^2 + \frac{1}{v}\left(F_i + \frac{2\pi}{\ell_0\ell} M_i\right)^2 \right]$$

$$+ \sum_x \frac{e_q}{e_v}\mu\eta\, Q_\mu Q_\mu + \sum_{x,i} i\frac{\mu\eta m v\ell}{2\pi}\left(\ell_0 A_0 Q_0 + \ell A_i Q_i\right). \tag{9.5}$$

For simplicity of presentation we shall introduce again our rescaled length variable in the time direction, $x^0 = vt$, so that $\nabla^2 = d_\mu d_\mu = (1/v)^2 d_t^2 + \sum_i d_i^2$, and we define dimensionless gauge fields

$$\begin{aligned} \mathcal{A}_0 &= \ell_0 A_0, \\ \mathcal{A}_i &= \ell A_i, \end{aligned} \tag{9.6}$$

and dimensionless (dual) field strengths

$$\mathcal{F}_\mu = \hat{K}_{\mu\nu} \mathcal{A}_\nu. \tag{9.7}$$

9.1. Superconductors

Although it is not the main focus of this book, let us first show how superconductivity emerges from the condensation of charges in a phase in which vortices are dilute, i.e. short-lived quantum fluctuations. We thus neglect M_μ and we sum only over the conserved integer currents Q_μ in (9.4) to obtain the ground state properties. We first solve the constraint $\hat{d}_\mu Q_\mu = 0$ by introducing the representation $Q_\mu = K_{\mu\nu} n_\nu$, with $d_\mu n_\mu = 0$, and we substitute the sum over the two independent integers encoded in $\{Q_\mu\}$ with the sum over the two independent integers in $\{n_\mu\}$. The Q-dependent part of (9.5) becomes then

$$S_{\text{top}}\left(Q_\mu, A_\mu\right) = \cdots + \sum_x \frac{e_q}{e_v}\mu\eta\, n_\mu\left(-\delta_{\mu\nu}\ell^2\nabla^2\right) n_\nu + i\frac{\mu\eta m v\ell}{2\pi} n_\mu \mathcal{F}_\mu. \tag{9.8}$$

We now turn the sum over $\{n_\mu\}$ into an integral by the Poisson formula

$$\sum_{n_\mu} f\left(n_\mu\right) = \sum_{k_\mu} \int dn_\mu f\left(n_\mu\right) e^{i2\pi n_\mu k_\mu}, \tag{9.9}$$

where the new integer link variables $\{k_\mu\}$ must satisfy $\hat{d}_\mu k_\mu = 0$ since only their transverse components can contribute. At this point we can evaluate the Gaussian integration over $\{n_\mu\}$ by using, as usual, the gauge fixing procedure explained in Chapter 3. The integers $\{k_\mu\}$ can be reabsorbed into the $\{M_\mu\}$ and play the same role of short-lived vortex excitations. The infrared-dominant contribution to the effective action is

$$S_{\text{eff}}\left(A_\mu\right) = \sum_x \frac{\ell}{\alpha\lambda} \mathcal{A}_\mu \left(\delta_{\mu\nu} - \frac{\hat{d}_\mu d_\nu}{\nabla^2}\right) \mathcal{A}_\nu + \dots, \qquad (9.10)$$

with

$$\lambda = \frac{16\pi^2\ell}{\alpha\mu\eta(mv\ell)^2}\frac{e_q}{e_v} = O\left(\frac{\lambda_L^3}{d\ell}\right). \qquad (9.11)$$

Gauge invariance is explicit, at the price of non-locality. The effective action becomes, instead, local in the Lorenz gauge $\Delta_\mu \mathcal{A}_\mu = 0$,

$$S_{\text{eff}}\left(A_\mu\right) = \sum_x \frac{\ell}{\alpha\lambda} \mathcal{A}_\mu \mathcal{A}_\mu + \cdots. \qquad (9.12)$$

This implies an induced current (rotated back to Minkowski space)

$$\mathbf{j} = \frac{v}{\alpha\lambda}\mathbf{A}, \qquad (9.13)$$

which, for a uniform charge distribution, is equivalent to the London equations

$$\begin{aligned}
\operatorname{rot}\mathbf{j} &= \frac{v}{\alpha\lambda}B, \\
\partial_t\mathbf{j} &= \frac{v}{\alpha\lambda}\mathbf{E}.
\end{aligned} \qquad (9.14)$$

This confirms that the charge condensate phase is a superconductor. The characteristic length of the superconductor is governed by the scale $\lambda = \lambda_L^3/d\ell$ in which all three relevant scales appear. In the vicinity of the SIT tricritical point, i.e. for $\ell/\lambda_L = O(1)$, it reduces to the Pearl length $\lambda_\perp = \lambda_L^2/d$ [5]. The characteristic energy scale of the superconductor is $v/\lambda = O\left((v/\lambda_L)(d\ell/\lambda_L^2)\right)$ and is thus typically smaller than the topological gap $O(v/\lambda_L)$.

9.2. Superinsulators

Let us now move to the dual phase, in which charges are short-lived excitations and magnetic topological defects M_μ condense. As we explained at the end of Chapter 7, the SIT can be viewed as driven by a diverging dielectric permittivity ε. To derive the properties of the insulating phase near the transition we can thus consider the limit $v = 1/\sqrt{\varepsilon} \ll 1$ in which magnetic fields in the effective action can be neglected with respect to electric fields, so that

$$S_{\text{top}}\left(M_\mu, A_\mu\right) = \sum_{x,i} \frac{1}{2e_{\text{eff}}^2} \left(\mathcal{F}_i + 2\pi M_i\right)^2, \tag{9.15}$$

which is nothing else than a deep non-relativistic version of Polyakov's compact QED action (5.29) in which only the electric fields survive and with an effective coupling constant

$$e_{\text{eff}}^2 = \frac{2\pi^2}{\mu} \frac{1}{\eta g} = e^2 \frac{\pi}{2\mu\eta} \frac{\lambda_L}{d} = e^2 O\left(\frac{\lambda_L}{d}\right). \tag{9.16}$$

This clearly shows again that making the film thinner is equivalent to increasing the effective strength of the Coulomb interaction.

The deep non-relativistic limit does not affect the main consequence of Polyakov's original idea [18] we presented in Sec. 5.3. The computation can be repeated step by step, with essentially the same result, notably that Cooper pairs cannot move anymore over large numbers of superconducting islands because Cooper pairs and Cooper holes are bound by a linear potential with string tension

$$\sigma = \frac{\sqrt{8}}{\ell_0 \ell} \frac{e_{\text{eff}}}{\pi} e^{-\frac{\pi^2}{e_{\text{eff}}^2} G_2(0)}, \tag{9.17}$$

where $G_2(0)$ is now the infrared-regularized 2D lattice Coulomb potential at coinciding points. This linear potential is due to a flux tube (string) of electric field connecting Cooper pairs and Cooper holes. When one pulls this string by, say, an external voltage, Cooper pairs and Cooper holes start moving apart but there comes a moment where it becomes energetically favourable for the system to pop out a Cooper-pair–Cooper-hole pair in some intermediate island and to form two short strings instead of a long one. As a consequence, only neutral states exist asymptotically in this phase of the system and the resistance becomes *infinite* since charges cannot move anymore. An applied voltage has the only consequence of creating neutral excitations in the system. This is an electric equivalent of the strong interaction that holds mesons and hadrons together, with Cooper pairs playing the role of *confined* quarks. We will dedicate an entire chapter to this analogy later.

There are, however two major differences that do arise as a consequence of the deep non-relativistic limit. The first concerns the shape of the monopoles that cause the linear confinement. Because of gauge invariance in the b_μ sector of the effective gauge theory (8.6) we have $d_\mu M_\mu = 0$. However, this does not imply $d_i M_i = 0$ and, indeed, the monopoles arise here only from the non-integer longitudinal spatial components M_i^L, $m = d_i M_i^L$, the transverse components M_i^T being reabsorbed into $\mathcal{F}_i$ due to the compactness of the gauge field. However, because of gauge invariance, these monopoles satisfy also $d_t M_0 = -m$. This means that we have also deeply non-relativistic monopoles, representing *instanton quantum tunneling events* [20] in which vortices on the film appear/disappear and their magnetic flux flows in/out isotropically in the four available spatial directions, as show in Fig. 9.1. This type of quantum tunneling can always happen with purely topological charges, in contrast to Noether charges implied by symmetry, which are always conserved. In the case of vortices this happens for Josephson vortices, which do not have a normal core and are characterized exclusively as a non-trivial gauge field configuration, contrary to Abrikosov vortices [5] (see also Chapter 11). These non-perturbative quantum tunneling events can be regarded as the 2D equivalent of the 1D phase slips [61–63] discussed in Sec. 5.2.

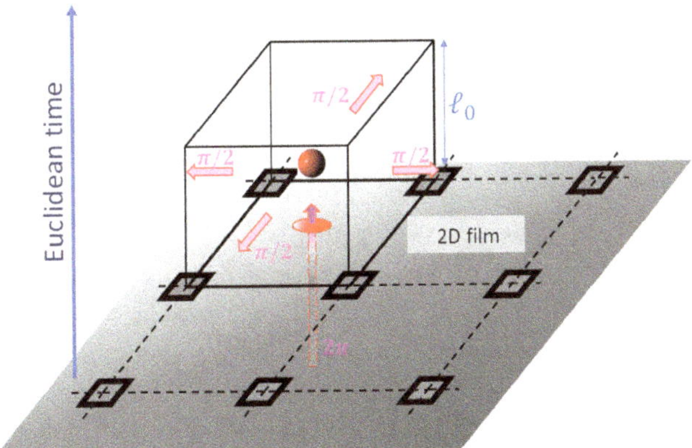

Fig. 9.1. A non-relativistic magnetic monopole instanton representing a quantum tunneling event in which a fundamental vortex of flux 2π at time t is divided up into four fluxes $\pi/2$ that flow out isotropically in the spatial directions. At the next instant $t + \ell_0$ there is no vortex anymore on the film. The condensate islands are represented schematically as in a regular array only for simplicity. From C. A Trugenberger, M. C. Diamantini, N. Poccia, F. S. Nogueira and V. M. Vinokur, Magnetic monopoles and superinsulation in Josephson junction arrays, *Quantum Reports* **2**, 388–399 (2020).

The second major difference is that the interaction of the monopoles near the SIT is logarithmic, $(e_{\text{eff}}^2/2\pi)\ln|\mathbf{x}|$, instead of an inverse linear power. This is because, in the deep non-relativistic limit, it is determined by the purely spatial inverse Laplacian ∇_2^2 instead of the full Laplacian in 3D Euclidean space-time,

$$Z = Z_0 \cdot Z_{\text{inst.}} = \int_{-\infty}^{+\infty} \mathcal{D}\mathcal{A}_\mu \; e^{-\frac{1}{2e_{\text{eff}}^2}\sum_{x,i}\mathcal{F}_i{}^2} \cdot \sum_{\{m\}} e^{-\frac{2\pi^2}{e_{\text{eff}}^2}\sum_x m\frac{1}{-\nabla_2^2}m}.$$

$$(9.18)$$

As a consequence, the lattice Coulomb potential $G_2(0)$ at coinciding points must be considered as regularized in the infrared by the usual screening effects of the plasma [47]. Contrary to the purely relativistic setting, in which monopoles are always in a plasma phase due to their weak inverse linear interaction, near the SIT they can undergo a confining quantum BKT transition [14] for sufficiently weak coupling constants. Of course, when monopoles are suppressed by their logarithmic interaction the confinement of electric charge they induce ceases. This quantum BKT transition is thus the SIT itself, as expected. Comparing with the exact BKT result derived in Sec. 5.1 we obtain the critical value $g_{\text{crit}} = (4\pi/\mu)(1/\eta)$: monopoles and linear confinement of charges can exist only for $g < g_{\text{crit}}$. This is in excellent agreement with the estimate obtained from the crude free energy argument for strings which would correspond to the string entropy value $\mu = 4\pi$.

We have derived that there is an infinite resistance due to magnetic monopole instantons at $T = 0$. However, every insulator with activated resistance has an infinite resistance at zero temperature. What is different here? To answer this question we will compute what happens to the monopoles when we raise the temperature. In particular, we will focus on the region $g \ll g_{\text{crit}}$ where the dielectric constant ε (at fixed screening length) is smaller and the system becomes near relativistic. Actually, for simplicity of calculation we will consider the fully relativistic model by taking $v = 1$. In this case, the monopole interaction is the usual Coulomb law in 3D Euclidean space-time, $e_{\text{eff}}^2/4\pi|x|$. As we derived in (7.4), at finite temperatures $T = 1/\beta$, and for spatial separations $0 \le |\mathbf{x}| < \beta$, this becomes $(e_{\text{eff}}^2/2\pi\beta)\ln(|\mathbf{x}|/\beta)$. Again, the monopole plasma undergoes a BKT transition [14], this time a thermal one, for high enough temperatures. Since the temperature is in the numerator of the potential, however, monopoles are free (plasma phase) at low temperatures and suppressed at high temperatures. In the high-temperature regime thus, the linear confinement of charges also ceases. Comparing with the BKT results in Sec. 5.1 we obtain the critical temperature for deconfinement as

$$T_{\text{dec}}(g) = \frac{e_{\text{eff}}^2}{2\pi^2 \ell} = \frac{1}{g} \frac{1}{\mu \eta \ell} \approx \frac{4e^2}{2\pi\varepsilon d}\left(0.08\frac{\varepsilon\lambda_L}{\ell}\right). \tag{9.19}$$

The second expression shows the dependence of the deconfinement temperature on the SIT coupling constant g: the smaller the coupling, the higher the deconfinement temperature. The final expression shows the dependence of the deconfinement temperature on the physical logarithmic Coulomb interaction of Cooper pairs, where we have taken $\eta \approx 1$ and $\mu \approx \ln 5$. We will use this expression later, when comparing with experiments.

This result clearly exposes what is different with respect to normal, activated insulators. There is a finite temperature range $0 < T < T_{\text{dec}}(g)$ in which charges are linearly confined and in this whole range the resistance is infinite, even at *finite temperatures*. This is a new state of matter. Since it is the exact dual of a superconductor we called it a *superinsulator* [6, 11, 16]. Superinsulation in thin films is thus caused by a plasma of magnetic monopole instantons.

The deconfinement BKT transition at $T = T_{\text{dec}}$ is a charge BKT transition dual to the usual vortex BKT transition out of the superconducting state [47]. This charge BKT transition has been experimentally detected in NbTiN films [15], as shown in Fig. 9.2. Here, the conductance in a NbTiN 10 nm thick film is plotted on a logarithmic scale as a function of the BKT scaling parameter $(T/T_{\text{CBKT}} - 1)^{-1/2}$, see Eq. (5.20). A linear plot indicates thus hyperactivation with the resistance diverging at T_{CBKT}: below this temperature superinsulation sets in.

9.3. Bose metals

Finally, for $\eta > 1$ we have an intermediate phase in which both charges and vortices are short-lived fluctuations, suppressed by a large gap and we can set effectively $Q_\mu = 0$ and $M_\mu = 0$ for the ground state properties. This gap arises from the topological interactions of charges and vortices and is quantified by the Chern–Simons mass [48] in their effective field theory, in which it plays the role of the UV cutoff. The bulk of the system is thus effectively frozen. Exactly as in superconductors, where the charge dynamics is massive, with a mass given by the inverse of the London penetration length, currents can flow only along the surface of the system. Here, however, the surface is a 1D edge and there is no superconductivity in 1D, because of

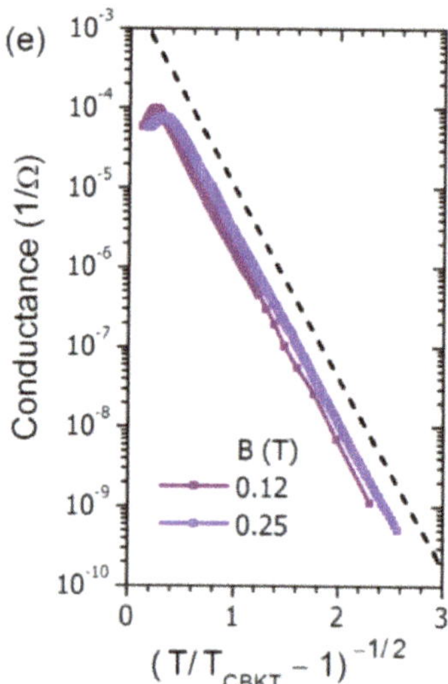

Fig. 9.2. The BKT temperature scaling of the conductance in a NbTiN film of 10 nm thickness. The charge deconfinement temperatures for the two value of the magnetic field are $T_{CBKT} = 30$ mK ($B = 0.12$ T) and $T_{CBKT} = 64$ mK ($B = 0.25$ T). From A. Yu. Mironov, D. M. Silevitch, T. Proslier, S. V. Postolova, M. V. Burdastyh, A. K. Gutakovskii, T. F. Rosenbaum, V. M. Vinokur and T. I. Baturina, Charge Berezinskii–Kosterlitz–Thouless transition in superconducting NbTiN films, *Scientific Reports* **8**, 4082 (2018), with permission by the authors. Creative Commons Attribution 4.0.

the resistance caused by both thermal and quantum phase slips [98]. The resulting behaviour is metallic, with a resistance saturating at finite values as $T \to 0$. We will dedicate the entire Chapter 10 to this phase of the system.

Chapter 10

Bose metals, a.k.a. bosonic topological insulators

Since Fisher first pointed out that the direct SIT transition point is a universal metal with exactly one quantum of sheet resistance $R_\square = R_Q = h/4e^2 = 2\pi/4e^2$ (in natural units) [9], it was predicted that, under certain conditions, this single point could open up to form a fully fledged metallic intermediate phase between superconductor and superinsulator [6]. This metal made out of bosons was subsequently called Bose metal [25, 26] and is known today also as bosonic topological insulator [97]. It was observed in numerous experiments on different materials [28–30] (for a review see [27]). In this chapter we will work in the continuum formulation in Minkowski space-time and we will show how the granular structure of the SIT [35] is typical of a bosonic topological insulator and self-organizes from the competition of two quantum BKT transitions.

In 2D, topological insulators (for a review see [99]) have the same effective field theory (6.25) [97] as the incompressible quantum Hall fluids presented in Chapter 6. The difference lies in the admitted K matrices; contrary to quantum Hall fluids, topological insulators are integer phases, with no topological order, and thus $|\det K| = 1$. In the simplest case $N = 2$ of two quasi-particles we have two possibilities, fermionic topological insulators, with K matrix

$$\begin{pmatrix} 0 & 1 \\ 1 & 1 \end{pmatrix} \tag{10.1}$$

and bosonic topological insulators, with K matrix

$$\begin{pmatrix} 0 & 1 \\ 1 & 0 \end{pmatrix} \tag{10.2}$$

up to $GL(2, \mathbb{Z})$ transformations [97]. We are interested here only in the latter, which is non-chiral, since the K matrix is symmetric, of even dimension. As a consequence there is no Hall current. But there is also no longitudinal current, since the bulk is frozen by an (infinite) gap. This state is thus a bosonic insulator, a topological bosonic insulator since the effective field theory and the gap are topological. Its effective field theory is given by the mixed Chern–Simons term (8.1)

$$S = \int d^3x \; \frac{1}{2\pi} a_\mu \epsilon^{\mu\alpha\nu} \partial_\alpha b_\nu + a_\mu Q^\mu + b_\mu M_\mu, \qquad (10.3)$$

which was posited to describe an emergent condensed matter state already in [6], albeit not yet under its modern name.

Unfortunately, the standard treatment of bosonic topological insulators is incomplete. The problem arises from the canonical structure of Chern–Simons theories, which are of first order in derivatives. When Maxwell terms are added as UV regulators to make the topological mass finite, the theory becomes of second order in derivatives, as usual. The topological limit of infinite mass involves thus a phase space reduction. It turns out that this phase space reduction does not commute with quantization [56]. The wave functional of pure Chern–Simons theories is different than the wave functional of Maxwell–Chern–Simons theories in the infinite mass limit. In particular, the former is non-normalizable while the second is. As a consequence, in physical applications to emergent condensed matter systems, Chern–Simons theories must always be considered as the infinite mass limit of Maxwell–Chern–Simons theories. This is of course plausible also from a physical point of view, since in such emergent condensed matter phases a UV cutoff is always present. When only one Chern–Simons gauge field is present this poses no problem. However, when two (or more) Chern–Simons terms are involved, as in (10.3), the topological mass is the product of the two possible Maxwell coupling constants, $m = e_q e_v / 2\pi$, as in (8.9). In this case, the "infinite mass limit" is not well specified if we do not fix the ratio $g = e_v / e_q$ of these two coupling constants. The action $S(g)$ of bosonic topological insulators depends thus on a hidden dimensionless parameter g which cannot be neglected. If the two gauge symmetries are compact, this hidden parameter g can lead to quantum phase transitions to superconductor and superinsulator phases by the condensation of the two quasi-particle types in (10.3), as was first shown in [6]. This is the essence of the physics of the SIT.

What are the consequences of this hidden coupling? We first focus on the electromagnetic response. This is given by the continuum version of (9.5) with $Q_\mu = 0$ and $M_\mu = 0$. Remembering that, near the tricritical point, we have $g \approx 1$, $\eta \approx 1$ and $\ell \approx \lambda_L$ and using $\lambda_L = O(d/e^2 g)$ we obtain (absorbing all numerical constants in the gauge fields)

$$S_{\text{eff}}(A_\mu) = \frac{1}{2} \frac{d}{e^2 g} \int d^3x \left(F_0^2 + \frac{1}{v^2} F_i^2 \right). \tag{10.4}$$

This describes an insulator, since the induced current $j^\mu \propto \partial_\nu F^{\nu\mu}$ (in the relativistic case $v = 1$ for simplicity of presentation) is proportional to the derivatives of the applied field and thus, in the bulk, both the longitudinal and the quantum Hall currents vanish, as anticipated. In the topological limit $m = e_q e_v / 2\pi \to \infty$ the bulk electromagnetic response vanishes altogether. However, (10.4) shows that, in this intermediate topological state, the charge e is renormalized to an effective charge $e_g = e\sqrt{g}$. While this is irrelevant for the frozen bulk, it becomes crucial for the boundary physics, as we now show.

To derive the boundary physics let us repeat the analysis of incompressible quantum Hall fluids at the end of Chapter 6, but with the mixed Chern–Simons action describing topological insulators this time,

$$S = \frac{1}{2\pi} \int d^3x \, a_\mu \epsilon^{\mu\nu\alpha} \partial_\nu b_\alpha + \frac{e_g}{2\pi} \int d^3x \, A_\mu \epsilon^{\mu\nu\alpha} \partial_\nu b_\alpha, \tag{10.5}$$

where we have included the electromagnetic coupling of the charge current $j^\mu = (1/2\pi)\epsilon^{\mu\alpha\nu}\partial_\alpha b_\nu$. As in the quantum Hall case, this action is not gauge-invariant under transformations that do not vanish on the boundary and one must add edge degrees of freedom so that gauge invariance of the complete theory is restored. In order to do this we consider the Weyl gauge $a_0 = 0$, $b_0 = 0$ and we set $a_i = \partial_i \lambda$ and $b_i = \partial_i \chi$. To restore full gauge invariance these fields have to be promoted to scalar edge degrees of freedom, with an edge action

$$S_{\text{edge}} = \frac{1}{4\pi} \int d^2x \, (\partial_0 \lambda \partial_s \chi + \partial_s \lambda \partial_0 \chi) + \frac{e_g}{2\pi} \int d^2x \, (A_0 \partial_s \chi - A_s \partial_0 \chi), \tag{10.6}$$

where s denotes the space coordinate along the one-dimensional edge. We now introduce the two new fields $\lambda = \xi + \eta$ and $\chi = \xi - \eta$. In terms of these, the edge action decouples,

$$S_{\text{edge}} = \frac{1}{2\pi} \int d^2x \; (\partial_0\xi\partial_s\xi - \partial_0\eta\partial_s\eta) + e_g \int d^2x \; A_0 \left(\frac{1}{2\pi}\partial_s\chi\right), \quad (10.7)$$

where we have taken, for simplicity's sake, $A_s = 0$ and from which we can identify

$$\rho = \frac{1}{2\pi}\partial_s\chi \qquad (10.8)$$

as the one-dimensional charge density of edge excitations.

As in the case of the quantum Hall effect, a non-universal dynamics for the edge modes is assumed to be generated by boundary effects. These contribute a Hamiltonian

$$H = \frac{1}{2\pi} \int ds \left[v_b \left(\partial_s\xi\right)^2 + v_b \left(\partial_s\eta\right)^2 \right], \qquad (10.9)$$

where v_b is the velocity of propagation of the edge modes along the boundary. Adding this term, the total edge action becomes

$$S_{\text{edge}} = \frac{1}{2\pi} \int d^2x \; \left[(\partial_0 - v\partial_s)\,\xi\partial_s\xi - (\partial_0 + v\partial_s)\,\eta\partial_s\eta \right]$$

$$+ e_g \int d^2x \; A_0 \left(\frac{1}{2\pi}\partial_s\chi\right). \qquad (10.10)$$

As compared to the quantum Hall case, we now have *two chiral bosons* circulating along the edge in the opposite directions (the different sign in the velocity term). Let us compute the equations of motion of this field theory model:

$$\left(\partial_0 - v\partial_s\right)\partial_s\xi = -\frac{2e_g}{2}E,$$

$$\left(\partial_0 + v\partial_s\right)\partial_s\eta = -\frac{2e_g}{2}E, \qquad (10.11)$$

where $E = \partial_s A_0$ is the applied electric field. These equations can be re-expressed in terms of λ and χ as

$$\partial_0\partial_s\lambda - v_b\partial_s^2\chi = -2e_gE,$$

$$\partial_0\partial_s\chi - v_b\partial_s^2\lambda = 0. \qquad (10.12)$$

In a static situation (all time derivatives vanishing) the first of these equations becomes

$$v_b \partial_s^2 \chi = 2e_g E.$$
(10.13)

Finally, using the identification (10.8) we rewrite this as

$$v_b \partial_s \rho = \frac{2e_g}{2\pi} E = \frac{2e_g}{2\pi} \partial_s V,$$
(10.14)

where we have used the voltage symbol V for A_0. The one-dimensional Cooper pair current being defined as $I = 2e_g v_b \rho$ we obtain the final equation

$$\frac{I}{V} = \frac{(2e_g)^2}{2\pi}.$$
(10.15)

The two chiral modes combine to a ballistic charge transport along the 1D edge. As a result, the system is a $T = 0$ Bose metal with sheet resistance

$$R_\square = \frac{1}{g} R_Q.$$
(10.16)

It is important to stress that this perfect conduction channel originates from symmetry protection under the symmetry $U(1) \times \mathbb{Z}_2^T$, where $U(1)$ is the global gauge symmetry and $\mathbb{Z}_2^T$ is time-reversal invariance [97, 100]. Insulators with this symmetry are characterized by a $\mathbb{Z}_2$ quantum number, where one possible value stands for trivial insulators and the other for topological insulators [97, 101]. All perturbations that do not change this quantum number cannot transform a topological insulator into a trivial one. Therefore, even in presence of impurities and disorder, the symmetry-protected charge carriers do not localize, even in 1D. Of course, to measure a finite resistance, there must be an energy release mechanism. This is provided by quantum phase slips of vortices, as described in Section 5.2, which are also mobile across the 1D edge.

For $g = 1$ we recover the original universal result of Fisher [9]. When the SIT point opens up to an intermediate metallic phase, however, the sheet resistance deviates from the resistance quantum, it becomes lower on the superconducting side, higher on the insulating side. This is exactly what is observed in experiments (see [27] for a review), as shown in Fig. 10.1 for a NbTiN film, with the SIT driven by a magnetic field [34]. The metallic saturation approaches constant values on both sides of the resistance quantum $R_Q = 6.45$ kΩ as the system is driven across the SIT.

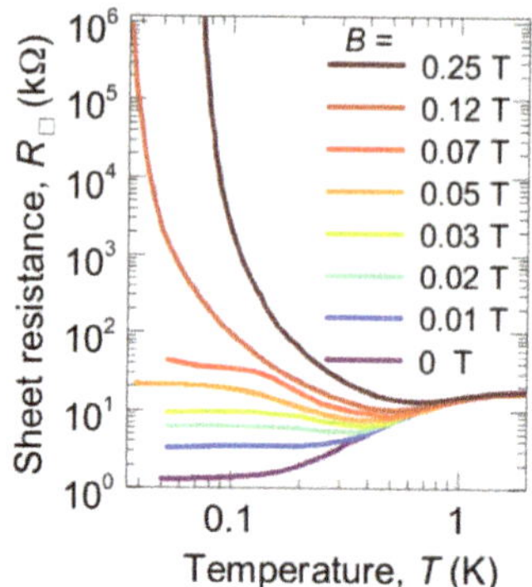

Fig. 10.1. Bose metal resistance saturation as $T \to 0$ in a NbTiN thin film in which the SIT is driven by an applied magnetic field. The metallic saturation approaches values on both sides of the resistance quantum $R_Q = 6.45$ kΩ as the system is driven across the SIT. From M. C. Diamantini, A. Yu. Mironov, S. V. Postolova, X. Liu, Z. Hao, D. M. Silevitch, Ya. Kopelevich, P. Kim, C. A. Trugenberger and V. M. Vinokur, Bosonic topological intermediate state in the superconductor–insulator transition, *Phys. Lett.* **A384**, 126570 (2020), ©Elsevier (2020).

In real systems, the topological gap is of course finite, although it can be large. Therefore, charge transport through the bulk should be activated at higher temperatures. This is also clearly observable in experiments. In Fig. 10.2 we display the sheet resistance of the NbTiN film as a function of $1/T$ for different values of the magnetic field [34]. The emergent minima perfectly fit the formula for two parallel resistors:

$$R_\square(T) = \frac{R_{\text{bulk}}(T)R_{\text{edge}}}{R_{\text{bulk}}(T) + R_{\text{edge}}}, \qquad (10.17)$$

where $R_{\text{edge}} = $ const. is the metallic resistance due to the edge modes and $R_{\text{bulk}} \propto \exp(\Delta_{\text{top}}/T)$ is the activated contribution of the bulk. A fit to this formula is shown in Fig. 10.2 by a solid line for a magnetic value of 0.05 T. Such emergent minima are considered the hallmark of topological insulators, see [102] and references therein.

Let us now derive how the hidden coupling g leads to phase transitions and the granular structure. Essentially we will show that the topological limit of infinite mass/gap corresponds to a functional generalization of the first Landau level of particles in a magnetic field. The coupling g plays a role similar to the filling fraction and drives transitions to other phases of this functional first Landau level. Let us consider the continuum, Minkowski space-time version of the effective action (8.5), for the moment without

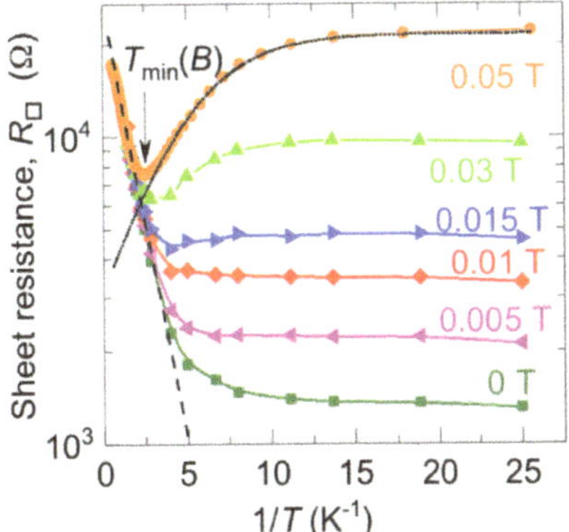

Fig. 10.2. Sheet resistance of a NbTiN film as a function of inverse temperature and for different values of applied magnetic field. The emergent minima perfectly fit the two-parallel-resistor formula for a metallic edge and an activated bulk, as shown by a solid line for the curve corresponding to a magnetic field 0.05 T. From M. C. Diamantini, A. Yu. Mironov, S. V. Postolova, X. Liu, Z. Hao, D. M. Silevitch, Ya. Kopelevich, P. Kim, C. A. Trugenberger and V. M. Vinokur, Bosonic topological intermediate state in the superconductor–insulator transition, *Phys. Lett.* **A384**, 126570 (2020), ©Elsevier (2020).

topological excitations,

$$S = \int dt\, d^2x \left(\frac{-v}{2g_0^2} f_0 f^0 + \frac{-1}{2g_0^2 v} f_i f^i + \frac{1}{2\pi} a_\mu \epsilon^{\mu\alpha\nu} \partial_\alpha b_\nu \right.$$
$$\left. + \frac{-v}{2e_0^2} g_0 g^0 + \frac{-1}{2e_0^2 v} g_i g^i \right), \tag{10.18}$$

where e_0^2 and g_0^2 are coupling constants with dimension [1/length] and v is the speed of light in the material. The charge current is $j^\mu = (1/2\pi) f^\mu$, with $f^\mu = \epsilon^{\mu\alpha\nu} \partial_\alpha b_\nu$ the dual field strength. When the gauge symmetries are taken as compact, the radii are 2π for both $U(1)_a$ and $U(1)_b$, respectively. To describe the vicinity of the SIT we will again consider the deep non-relativistic limit $v \ll 1$ and keep only electric fields in the action,

$$S = \int dt\, d^2x \left(\frac{-1}{2e_q^2} g_i g^i + \frac{q}{2\pi} a_\mu \epsilon^{\mu\alpha\nu} \partial_\alpha b_\nu + \frac{-1}{2e_v^2} f_i f^i \right), \tag{10.19}$$

where we have reabsorbed the factor v to obtain couplings with dimension [energy], $e_q^2 = e_0^2 v$, $e_v^2 = g_0^2 v$.

We shall now quantize the model (10.19) in the functional Schrödinger picture. As usual for a gauge theory, the gauge components a^0 and b^0

are not dynamical fields, since they never appear in the action with time derivatives. They are Lagrange multipliers, whose associated Gauss law constraints implement gauge invariance. They can be set to zero, $a^0 = 0$ and $b^0 = 0$, after imposing the corresponding Gauss law constraints. This is called the Weyl gauge. The two canonical momenta conjugate to the canonical variables a^i and b^i are:

$$
\begin{aligned}
\mathcal{P}_a^i &= \frac{\delta \mathcal{L}}{\delta(\partial_0 a^i)} = \frac{1}{e_q^2} g^{0i} + \frac{1}{4\pi} \epsilon^{ij} b^j, \\
\mathcal{P}_b^i &= \frac{\delta \mathcal{L}}{\delta(\partial_0 b^i)} = \frac{1}{e_v^2} f^{0i} + \frac{1}{4\pi} \epsilon^{ij} a^j.
\end{aligned}
\tag{10.20}
$$

They are realized as functional derivatives,

$$
\mathcal{P}_a^i = -i \frac{\delta}{\delta a^i}, \qquad \mathcal{P}_b^i = -i \frac{\delta}{\delta b^i}.
\tag{10.21}
$$

The Hamiltonian density, when written in canonical variables, takes the form

$$
\mathcal{H} = \frac{e_q^2}{2} \left(\Pi_a^i \right)^2 + \frac{e_v^2}{2} \left(\Pi_b^i \right)^2,
\tag{10.22}
$$

where

$$
\begin{aligned}
\Pi_a^i &= \mathcal{P}_a^i - \frac{1}{4\pi} \epsilon^{ij} b^j, \\
\Pi_b^i &= \mathcal{P}_b^i - \frac{1}{4\pi} \epsilon^{ij} a^j
\end{aligned}
\tag{10.23}
$$

are the kinetic momenta. Due to the Chern–Simons term, the kinetic momenta do not commute,

$$
\left[\Pi_a^i(\mathbf{x}), \Pi_b^j(\mathbf{y}) \right] = -i \frac{1}{2\pi} \epsilon^{ij} \delta^2(\mathbf{x} - \mathbf{y}),
\tag{10.24}
$$

which is tantamount to the presence of a non-trivial functional gauge connection

$$
\begin{aligned}
\mathcal{A}_a^i(a, b) &= \frac{1}{4\pi} \epsilon^{ij} b_j, \\
\mathcal{A}_b^i(a, b) &= \frac{1}{4\pi} \epsilon^{ij} a_j.
\end{aligned}
\tag{10.25}
$$

Canonical momenta are not functional gauge invariant quantities, since they can be traded with the connection by a functional gauge transformation of the wave functionals,

$$\Psi\left[a^i, b^i\right] \to e^{i\frac{1}{4\pi}\Lambda(a,b)}\ \Psi\left[a^i, b^i\right],$$

$$\epsilon^{ij}b^j \to \epsilon^{ij}b^j + \frac{\delta}{\delta a^i}\Lambda\left(a, b\right),$$

$$\epsilon^{ij}a^j \to \epsilon^{ij}a^j + \frac{\delta}{\delta b^i}\Lambda\left(a, b\right),$$

$$\partial_i \frac{\delta}{\delta a^i}\Lambda\left(a^i, b^i\right) = 0, \quad \partial_i \frac{\delta}{\delta b^i}\Lambda\left(a^i, b^i\right) = 0,$$

(10.26)

where the last conditions are required to respect traditional gauge invariance, encoded in the Gauss law constraints, see below. Only the electric fields Π_a^i and the charge currents Π_b^i are well-defined gauge invariant quantities. The functional connection (10.25) is not pure gauge. The quantity

$$\mathcal{B}^{ij}(\mathbf{x} - \mathbf{y}) = \frac{\delta}{\delta a^i(\mathbf{x})}\mathcal{A}_b^j\left(a(\mathbf{y}), b(\mathbf{y})\right) - \frac{\delta}{\delta b^j(\mathbf{y})}\mathcal{A}_a^i\left(a(\mathbf{x}), b(\mathbf{y})\right)$$

$$= -\frac{1}{2\pi}\epsilon^{ij}\delta^2(\mathbf{x} - \mathbf{y})$$

(10.27)

plays the role of a functional uniform magnetic field and appears as the commutator of kinetic momenta, exactly as in the traditional Landau problem of electrons in an external magnetic field. This shows that the Chern–Simons term plays the role of a functional magnetic field, i.e. of a non-trivial curvature in configuration space.

Exactly as in the standard problem of Landau levels we can define lowering and raising operators (note that from now on in this section curly $\mathcal{A}$s denote raising and lowering operators)

$$\mathcal{A}^i = \sqrt{\frac{\pi}{e_q e_v}}\left(e_q\Pi_a^i - ie_v\epsilon^{ij}\Pi_b^j\right),$$

$$\mathcal{A}^{i\dagger} = \sqrt{\frac{\pi}{e_q e_v}}\left(e_q\Pi_a^i + ie_v\epsilon^{ij}\Pi_b^j\right),$$

(10.28)

with commutation relation

$$\left[\mathcal{A}^i(\mathbf{x}), \mathcal{A}^{j\dagger}(\mathbf{y})\right] = \delta^{ij}\ \delta^2(\mathbf{x} - \mathbf{y}).$$

(10.29)

In terms of these, the Hamiltonian takes the familiar form

$$H = \Delta_{\text{top}}\sum_i \int d^2\mathbf{x}\left(\mathcal{A}^{i\dagger}(\mathbf{x})\mathcal{A}^i(\mathbf{x}) + \frac{1}{2}\ \delta^{ii}\ \delta^2(\mathbf{0})\right),$$

(10.30)

where the second term represents the infinite ground state energy that has to be subtracted and $\Delta_{\text{top}} = e_q e_v / 2\pi$ is the topological energy gap. Finally, the Gauss law operators implementing gauge invariance are the constraints associated with the Lagrange multipliers a_0 and b_0,

$$
\begin{aligned}
G_a &\equiv \partial_i \mathcal{P}_a^i + \frac{1}{4\pi} \partial_i \epsilon^{ij} b^j, \\
G_b &\equiv \partial_i \mathcal{P}_b^i + \frac{1}{4\pi} \partial_i \epsilon^{ij} a^j.
\end{aligned}
\tag{10.31}
$$

At the quantum level, these constraints must be imposed as conditions on physical states:

$$
G_a \Psi[a^i, b^i] = 0, \quad G_b \Psi[a^i, b^i] = 0.
\tag{10.32}
$$

The ground state wave functional Ψ_0 is thus given by the symmetric gauge functional first Landau level, defined by $\mathcal{A}^i(\mathbf{x}) \Psi_0[a^i, b^i] = 0$, subject to the gauge constraints (10.32).

Localized excited states of unit norm and energy Δ_{top} are created by the operators

$$
\mathcal{A}_f^\dagger = \int d^2\mathbf{x}\, f(\mathbf{x} - \mathbf{x}_0)\, \hat{n}^i \mathcal{A}^{i\dagger},
$$

$$
\int d^2\mathbf{x}\, f^2(\mathbf{x} - \mathbf{x}_0) = 1,
\tag{10.33}
$$

with commutation relations

$$
\left[\mathcal{A}_f, \mathcal{A}_f^\dagger \right] = 1,
$$

$$
\left[H, \mathcal{A}_f^\dagger \right] = \Delta_{\text{top}} \mathcal{A}_f^\dagger.
\tag{10.34}
$$

Their energy does not depend on the form factor $f(\mathbf{x} - \mathbf{x}_0)$, as long as it satisfies the normalization condition (10.33).

We have obtained the anticipated result: the topological limit $\Delta_{\text{top}} \to \infty$ corresponds to decoupling the higher functional Landau levels. From now on we shall thus focus on the lowest Landau level only. Following [48, 103] we write the ground state wave functional as the product of a phase and a contribution that depends only on the transverse components of the two dynamical variables, a_T^i and b_T^i:

$$\Psi_0[a^i, b^i] = e^{i\chi(a^i, b^i)}\, \Phi(a_T^i, b_T^i),$$

$$\chi[a^i, b^i] = \frac{1}{4\pi} \int d^2\mathbf{x} \left(b\frac{\partial_i}{\Delta}a^i + a\frac{\partial_i}{\Delta}b^i \right), \tag{10.35}$$

$$\Phi[a_T^i, b_T^i] = \exp\frac{-1}{4\pi} \int d^2\mathbf{x} \left(\frac{e_v}{e_q}(a_T^i)^2 + \frac{e_q}{e_v}(b_T^i)^2 \right),$$

where $a = \epsilon^{ij}\partial_i a^j$, $b = \epsilon^{ij}\partial_i b^j$, $\Delta = \partial_i\partial_i$ is the purely spatial Laplacian, and $a_T^i = P^{ij}a^j$, $b_T^i = P^{ij}b^j$, with the projector P^{ij} onto the transverse part of the gauge fields given by $P^{ij} = \left(\delta^{ij} - \partial^i\partial^j/\Delta\right)$. Using the Hodge decomposition for the spatial components of the two gauge fields a^i and b^i:

$$a^i = \partial_i\xi + \epsilon^{ij}\partial_j\phi,$$
$$b^i = \partial_i\lambda + \epsilon^{ij}\partial_j\psi, \tag{10.36}$$

we can rewrite $\Psi_0[a^i, b^i]$ as:

$$\Psi_0[a^i, b^i] = e^{\frac{i}{4\pi} \int d^2\mathbf{x}\, (\psi\Delta\xi + \phi\Delta\lambda)} e^{\frac{-1}{4\pi} \int d^2\mathbf{x}\, \left(g(\partial_i\phi)^2 + \frac{1}{g}(\partial_i\psi)^2\right)}, \tag{10.37}$$

where $g = e_v/e_q$ is our hidden coupling constant. This shows that, even after having decoupled the higher Landau levels in the $\Delta_{\text{top}} \to \infty$ limit, the ground state wave functional still depends on the ratio g. As always in Chern–Simons gauge theories, gauge invariance is realized with a 1-cocycle [40], which manifests itself in the phase in (10.35) and (10.37). This phase can be expressed also as

$$e^{i\chi(a^i, b^i)} = e^{\frac{i}{2} \int d^2\mathbf{x}\left(\lambda\phi^0 + \xi j^0\right)}, \tag{10.38}$$

where $j^0 = (1/2\pi)\epsilon^{ij}\partial_i b_j$ and $\phi^0 = (1/2\pi)\epsilon^{ij}\partial_i a_j$ are the charge and vortex densities, respectively.

Two possibilities have to be considered. In the simplest case both gauge symmetries are non-compact, with gauge group $\mathbb{R}$. In this case, neither charges nor vortices are quantized and ground state connected quantum correlation functions of their densities are given by

$$\langle j^0(\mathbf{x})j^0(\mathbf{y})\rangle_c = \frac{1}{4\pi^2}\frac{1}{Z_\psi} \int \mathcal{D}\psi\, \Delta\psi(\mathbf{x})\Delta\psi(\mathbf{y})\, e^{\frac{-1}{2\pi g} \int d^2\mathbf{x}\, (\partial_i\psi)^2}, \tag{10.39}$$

$$Z_\psi = \int \mathcal{D}\psi\, e^{\frac{-1}{2\pi g} \int d^2\mathbf{x}\, (\partial_i\psi)^2},$$

with an analogous expression for vortices in terms of the field ϕ and with $g \to 1/g$. This gives

$$\langle j^0(\mathbf{x})j^0(\mathbf{y})\rangle_c = \pi g \, \Delta \delta^2(\mathbf{x}-\mathbf{y}), \quad \langle \phi^0(\mathbf{x})\phi^0(\mathbf{y})\rangle_c = \frac{\pi}{g} \, \Delta\delta^2(\mathbf{x}-\mathbf{y}),$$

$$(10.40)$$

which represent short-range, screened correlations. The vanishing screening length implied by these expressions is a consequence of the projection to the lowest Landau level. In the general case (10.18) the real part of the ground state wave functional (10.35) is modified to

$$\Phi[a_T^i, b_T^i] = e^{-\int d^2\mathbf{x} \left(\frac{1}{2e_q^2} a_T^i \sqrt{\Delta_{\text{top}}^2 - v^2\Delta} \, a_T^i + \frac{1}{2e_v^2} b_T^i \sqrt{\Delta_{\text{top}}^2 - v^2\Delta} \, b_T^i \right)}, \quad (10.41)$$

which shows that both charge and vortex densities are correlated on a typical length $\lambda_{\text{corr}} = v/\Delta_{\text{top}}$. The topological limit is analogous to taking the limit of infinite cyclotron frequency in the Landau problem, in which the magnetic length vanishes.

Things change when the two gauge symmetries are compact U(1) instead of non-compact $\mathbb{R}$. In this case, the fields ξ and λ in (10.36) are angles and the identity $\epsilon^{ij}\partial_i\partial_j\theta = 2\pi\delta^2(\mathbf{x})$ for an angle $\theta \in [0, 2\pi]$ implies the existence of quantized vortices and point charges. As we now show, in this U(1) $\times$ U(1) case, the fields ξ and λ become new dynamical fields embedding the dynamics of these additional degrees of freedom.

We start by noting that, in case of compact gauge symmetries, the cocycle is changed to

$$e^{i\chi(a^i, b^i)} = e^{\frac{i}{2}\int d^2\mathbf{x}\left(\lambda\phi^0 + \xi j^0\right)} e^{\frac{i}{2}\sum_i \int d^2\mathbf{x}\, [\xi(\mathbf{x})N(\mathbf{x},\mathbf{x}_i,N_i) + \lambda(\mathbf{x})\Phi(\mathbf{x},\mathbf{x}_i,\Phi_i)]},$$

$$(10.42)$$

where

$$N(\mathbf{x}, \mathbf{x}_i, N_i) = N_i \, \delta^2(\mathbf{x}-\mathbf{x}_i),$$
$$\Phi(\mathbf{x}, \mathbf{x}_i, \Phi_i) = \Phi_i \, \delta^2(\mathbf{x}-\mathbf{x}_i) \qquad (10.43)$$

represent the additional point particle and vortex degrees of freedom. The integers N_i and Φ_i encode the particle and vortex numbers, while $\mathbf{x}_i$ denotes their locations. These charges and vortices can condense.

To see how this is reflected in the functional Schrödinger picture, let us consider a mixed state for the vortex sector, arising from entanglement in the charge sector between charge observables $N(\mathbf{x}, \mathbf{x}_i, N_i)$ and $\psi(\mathbf{x})$ at different points. The expectation values of vortex sector operators $O(\phi, \xi)$ in this mixed state are given by

$$\langle O \rangle = \frac{1}{Z} \int \mathcal{D}\phi \mathcal{D}\xi \mathcal{D}\psi \; \sum_N \frac{z^N}{N!} \sum_{\mathbf{x}_1 \ldots \mathbf{x}_N} \sum_{N_1 \ldots N_N = \pm 1} O(\phi, \xi)$$

$$\times e^{\frac{i}{2} \sum_i \int d^2\mathbf{x} \; \xi(\mathbf{x}) N(\mathbf{x}, \mathbf{x}_i, N_i) + i \int d^2\mathbf{x} \; \frac{1}{4\pi} \xi \Delta \psi}$$

$$\times e^{\frac{-1}{2\pi} \int d^2\mathbf{x} \left(g(\partial_i \phi)^2 + \frac{1}{4g} (\partial_i \psi)^2 \right)}. \tag{10.44}$$

where Z is the normalization factor and we have traced over the charge observables ψ and N, ψ denoting thus the difference $\psi = \psi_{\text{bra}} - \psi_{\text{ket}}$ between bra and ket states. We have also used the approximation in which only interferences between point charge states differing by one charge unit are taken into account. The parameter z governs the entanglement. For $z \to 0$ we have a highly entangled state of charge, since configurations with independent charge degrees of freedom at every point ($N \gg 1$) are strongly suppressed, for $z \to 1$ charges are liberated as independent degrees of freedom. At this stage both the integration over the transverse field ψ and the summation over charge interference configurations can be done explicitly [19]. For a Cooper pairs state, with charge unit 2 and N_i being thus multiples of 2, the result is

$$\langle O \rangle = \frac{1}{Z} \int \mathcal{D}\phi \mathcal{D}\xi \; O(\phi, \xi) \; e^{-\int d^2\mathbf{x} \left(\frac{g}{2\pi} (\partial_i \phi)^2 + \frac{g}{4\pi} (\partial_i \xi)^2 - 2z \cos \xi \right)}. \tag{10.45}$$

Vortex sector observables in the entangled mixed state are thus determined by the classical partition function of the 2D sine–Gordon model or, equivalently, the 2D XY model [47] which we discussed extensively in Sec. 5.1. While the original ϕ field plays the role of the spin wave field, the new dynamical sine–Gordon field ξ describes the vortex dynamics. We can now also use the classical results on the 2D sine–Gordon (or XY) renormalization group flow [59] here.

The 2D XY model undergoes the famed BKT [14] phase transition. The self-dual value $g = 1$ separates a weak coupling phase for $g < 1$ from a strong coupling phase for $g > 1$. In both phases the coupling constant g flows to large values in the IR limit. Of course, only the $g < 1$ (left-hand side) of the flow diagram (Fig. 5.1) is relevant here, since the other, $1/g < 1$ half is dominated by vortex confinement, as discussed below. In the weak coupling phase, $z \to 0$ in the IR limit; while in the strong coupling phase, z increases in the IR limit. The renormalization flow depends on a constant C, which represents a particular combination of the initial conditions for the flow.

In the case $C > 0$, of interest here, there is no IR fixed point in the strong coupling phase, while the BKT critical point $g_{\text{crit}} = 1/(1+C)$ is a *confining* IR fixed point. At this point $z = 0$ and charge cannot fluctuate independently at differing space points. There is only one global charge degree of freedom. However, since configurations with a unit charge at every space point imply infinite energy, only neutral configurations survive as asymptotic states in the IR limit. In this state with infinite charge correlations, the conjugated phase ξ becomes a massless field. Note, however, that this does not describe fluctuations around a fixed, global phase. It is truly the whole phase degree of freedom which becomes massless, representing a vortex condensate: this is the superinsulator phase. Of course, one can now repeat verbatim the same analysis for a mixed state arising from entanglement in the vortex sector, which describes a charge condensate with vortex confinement. The dual critical point at $g_{\text{crit}} = (1 + C)$ represents the dual, superconducting phase. Combining the two halves $g < 1$ and $1/g < 1$ of the flow diagram we obtain a third fixed point at $g_{\text{crit}} = 1$, $z_{\text{crit}} > 0$ for $C < 0$. This describes the Bose metal.

As anticipated, the hidden coupling constant g leads to important consequences that are normally overlooked in the standard treatment of bosonic topological insulators. Let us now focus on the implications of the two quantum BKT transitions between the intermediate Bose metal and the superconductor and superinsulator phases. The intermediate Bose metal corresponds to both the high-temperature regimes $T_{\text{eff}} = g > g_{\text{crit}}$ for the quantum vortex condensation and $T_{\text{eff}} = 1/g > 1/g_{\text{crit}}$ for the quantum charge condensation. As we have derived in Sec. 5.1, see (5.19), in this regime we have exponential correlation functions

$$\langle e^{i\xi(\mathbf{x})} e^{-i\xi(\mathbf{y})} \rangle \propto \exp\left(-\frac{|\mathbf{x} - \mathbf{y}|}{\xi_{\text{XY}}(g)} \right),$$

$$\langle e^{i\lambda(\mathbf{x})} e^{-i\lambda(\mathbf{y})} \rangle \propto \exp\left(-\frac{|\mathbf{x} - \mathbf{y}|}{\xi_{\text{XY}}(1/g)} \right), \tag{10.46}$$

where the BKT correlation length $\xi_{\text{XY}}(g)$ diverges for g decreasing and approaching the critical value for vortex condensation and $\xi_{\text{XY}}(1/g)$ diverges for g increasing and approaching the critical value for charge condensation. The Bose metal, a.k.a bosonic topological insulator, is thus characterized by length scales $\xi_{\text{XY}}(g)$ and $\xi_{\text{XY}}(1/g)$ determining the typical length of superinsulating and superconducting fluctuations in this state. The scales become identical at the self-dual point $g = 1$. The Bose metal/bosonic topological insulator state arises thus from the competition of two quantum BKT transitions for charges and vortices which are both kept out of condensate by strong quantum fluctuations.

To establish what is the ground state that forms as a result of this competition we follow [104]. Since both charges and vortices are out of condensate, the infrared-dominant interactions are the mutual statistics interactions encoding the Aharonov–Bohm and Aharonov–Casher phases among charges and vortices and, as pointed out by Wilczek [55], these can be represented locally by introducing two fictitious gauge fields a_μ and b_μ with mixed Chern–Simons interaction as in (8.1). To proceed we use the technique of [105] and we consider vortices of the two kinds as living on two fictitious planes that we will solder together at the end. Let us begin by considering only up vortices. In this case the time-reversal symmetry is broken, it will be restored later when we combine the two chiralities. We shall now adopt the usual mean field approximation of anyon physics (for a review see [79]) by replacing the vortex distribution with a uniform magnetic field $\mathcal{B}_\Phi = 2eN_\Phi\Phi_0/A$ felt by the Cooper pairs, where N_Φ is the number of vortices of winding $\Phi_0 = \pi/e$ and A the area of the sample, and the charge distribution with a uniform "magnetic field" $\mathcal{B}_Q = \Phi_0 N_Q 2e/A$ felt by the vortices via the Magnus force, the dual of the Lorentz force, see e.g. [106] (N_Q is the number of charges). The Magnus force is the force on a vortex with a velocity $\mathbf{v}$ (with respect to the background). In two dimensions (2D) it takes the form

$$F_{\mathrm{M}}^i = \kappa\epsilon^{ij}v^j\rho, \qquad (10.47)$$

where ρ is the fluid density and κ is the vorticity, the winding of the vortex quantized in multiples of 2π. When the fluid is charged (with charge $2e$) it can be written as

$$F_{\mathrm{M}}^i = \Phi\epsilon^{ij}v^j 2e\rho, \qquad (10.48)$$

where Φ is the winding normalized in units of Φ_0. This is the exact dual of the usual Lorentz force $F_{\mathrm{L}}^i = 2e\epsilon^{ij}v^j B$, from which we evince that the charge density plays for vortices exactly the same role as a magnetic field for charges.

At the self-dual point, where $N_Q = N_\Phi$, the two "magnetic fields" $\mathcal{B}_Q$ and $\mathcal{B}_\Phi$ become identical, $\mathcal{B}_Q = \mathcal{B}_\Phi = \mathcal{B}$. At this point we can use the known results for the quantum state of two flavours of particles with Coulomb interactions and mutual statistics in a magnetic field [55, 107] to write down the wave function of the incompressible quantum fluid of charges and vortices at the self-dual point. In the general case it is

$$\Psi_{m_1,m_2,n}\left(\{z_i\},\{w_i\}\right) = \prod_{i,j}(z_i - z_j)^{m_1}\prod_{i,j}(w_i - w_j)^{m_2}$$

$$\times \prod_{i,j}(z_i - w_j)^n \; e^{-\frac{B}{4}\sum_i\left(|z_i|^2 + |w_i|^2\right)}, \quad (10.49)$$

where z_i denote the coordinates of the Cooper pairs and w_i those of vortices and m_1 and m_2 are even integers to guarantee the bosonic statistics of Cooper pairs and vortices. The total filling fraction (summing over both flavours) ν and the torus degeneracy D_T are given by [55]

$$\nu = \frac{m_1 + m_2 - 2n}{m_1 m_2 - n^2},$$

$$D_T = |m_1 m_2 - n^2|. \quad (10.50)$$

Restricting to the integer case (for both flavours) requires setting $m_1 = m_2 = 0$ and $n = 1$. In this case there is no topological order, encoded in a non-trivial torus degeneracy and the mutual statistics is also integer. It turns out, however, that the wave function in the integer case cannot be obtained by simply setting $m_1 = 0$ and $m_2 = 0$ in the above formula since the resulting state would be unstable with respect to phase separation [108]. The correct wave function of the topological incompressible fluid of the two-boson system (charges and up vortices) at the SIT self-dual point is [108]

$$\Psi\left(\{z_i\},\{w_i\}\right) = \prod_{i<j}|z_i - z_j|\prod_{i<j}|w_i - w_j|\prod_{i,j}\frac{(z_i - w_j)}{|z_i - w_j|}\; e^{-\frac{B}{4}\sum_i\left(|z_i|^2 + |w_i|^2\right)}.$$

$$(10.51)$$

Exactly as in the analogous quantum Hall effect situation [94], small deviations from the self-dual point will cause the system to produce small quantities of quasi-particle and quasi-hole excitations of the two kinds. It should come as no surprise that, for an integer topological state, these excitations are charges and vortices themselves [55]. On the superconducting side there will be an overabundance of charges, with an overabundance of vortices on the insulating side. And, as pointed out in [94], for sufficiently small densities near the self-dual point, these excitations behave as classical particles with fixed guiding centers in the respective magnetic fields. Since both have long-range Coulomb interactions, they will form a Wigner crystal (for a review see [109]) for sufficiently low temperatures.

The exact same wave function can be formulated for the down vortices on the second fictitious plane, with the only difference that it has the opposite chirality. Soldering the two fictitious planes means taking their "singlet quantum superposition", i.e. buidling a quantum superposition of conjugate up and down vortex configurations at the same locations with a relative (-1) sign [110]. This wave function describes the simplest bosonic topological insulator [97], with a long-distance effective Lagrangian

$$\mathcal{L} = \frac{1}{2\pi} a_\mu \epsilon^{\mu\alpha\nu} \partial_\alpha b_\nu, \tag{10.52}$$

expressed in terms of two effective vector and pseudovector gauge fields a_μ and b_μ [55, 108], as first obtained in [6]. The $U(1) \times \mathbb{Z}_2^T$ symmetry-protected edge modes of this model support an electric current with exactly one resistance quantum [34].

The ground state around the self-dual point consists thus primarily of a composite topological incompressible quantum fluid of Cooper pairs and vortices. Excess charges and vortices (with respect to integer filling) organize themselves as a Wigner crystal of guiding centres of singlet quantum superpositions of the two possible rotation chiralities. As shown above, when sufficient charge excitations are created, these will condense by a quantum BKT transition and a superconductor forms. The same happens with vortices on the other side of the self-dual point, with condensation giving rise to a superinsulator.

For $\eta > 1$, therefore, the observed granular structure [35, 88, 89] consists of a regular Wigner crystal of Cooper pairs and vortices. Since a Wigner crystal is insulating, electric conduction remains assured by the edge modes of the topological state. The only difference with respect to the quantum Hall effect case is that the creation of a small amount of quasi-particles slightly modifies the two "magnetic fields" felt by the two flavours of excitations. This effect is balanced by effective couplings e' and Φ_0' so that

$$e' N_\Phi' \Phi_0 = e N_Q' \Phi_0', \tag{10.53}$$

which can be rewritten as

$$\frac{e'}{e} = \frac{N_Q' \Phi_0'}{N_\Phi' \Phi_0}. \tag{10.54}$$

If we require that the Dirac quantization condition (i.e. the integer mutual statistics) is maintained we will thus have the same topological quantum state but with an effective, renormalized coupling constant

$$e \to e' = e\sqrt{g},$$

$$g = \frac{N_Q'}{N_\Phi'},$$
$$(10.55)$$

where we have identified the ratio of quasi-particles with the SIT parameter g. This effect leads to a renormalized resistance quantum

$$R_\Box = \frac{R_Q}{g} \qquad (10.56)$$

around the self-dual point, the same result we have already obtained above from the long-distance field theory.

For $\eta < 1$, instead there is no intermediate topological state. In this case, there is a finite range around the self-dual point in which charge and vortex condensates coexist in a first-order quantum transition [6,34]. In this case we have a "liquid" granular structure consisting of coexisting bubbles of superconducting and superinsulating condensates. This is the competing phase separation state already discussed in [108]. Finally, BKT physics implies the existence of a typical length scale also at low temperatures, in the condensate phase. This determines the typical distance of the logarithmically confined pairs of vortices (in the charge condensate) and pairs of charges (in the vortex condensate) [111]. This scale is responsible for the granular structure within the superconducting and superinsulating phases. In all cases the granular structure is self-organized and does not depend on the presence of structural disorder.

BKT physics implies logarithmic confinement [47]. In the superconducting phase this is the complete description. The magnetic monopole instantons interpolating between different vortex numbers, however, turn this into linear confinement of charges in the superinsulating phase, as we have shown in Sec. 9.2. The granularity scale ℓ determines also the string tension of this linear potential and sets thus also the scale of linearly bound pairs of charges, see (9.17). The BKT renormalization flux implies that z increases in the UV, i.e. charges become more and more free when looking at smaller and smaller scales. As we hit the scale $O(\ell)$, there is a regime where charges essentially do not feel any potential anymore, as we will derive in detail below. This phenomenon is called *asymptotic freedom* and is the paradigmatic feature of non-Abelian gauge theories such as QCD [24]. Contrary to QCD, however, where asymptotic freedom is characteristic of the UV gauge theory fixed point, here we obtain the dual phenomenon, in which it is the IR string theory fixed point which is perturbative ($z \ll 1$), while the UV regime is non-perturbative ($z \to 1$). Superinsulators constitute thus an "exactly solvable" U(1) model of confinement in which the usually non-accessible region of QCD becomes perturbative.

Quantum critical points are characterized by scaling [68]. In the vicinity of a quantum critical point B_{cr} driven by a coupling constant B, there is a characteristic length scale ξ scaling with a critical exponent ν as $\xi \propto |B - B_{\mathrm{cr}}|^{-\nu}$ and a characteristic frequency Ω scaling as $\Omega \propto \xi^{-z}$, where z is the dynamical critical exponent. Furthermore, near zero, the temperature T should scale as the frequency and thus $T \propto \xi^{-z}$. Using the simple fact that

$$|B - B_{\mathrm{cr}}| \propto \xi^{\frac{-1}{\nu}} \propto T^{\frac{1}{z\nu}}, \qquad (10.57)$$

we obtain the result that the scaling variable $s = |B - B_{\mathrm{cr}}|/T^{1/\nu z}$ is independent of B and T. In the vicinity of the quantum critical point we can thus substitute two variables B and T by one single universal scaling variable s. In terms of this scaling variable we should thus obtain that all quantities, when plotted in terms of s near criticality, should display two branches, one for each phase. All diverse measurements depending on B and T collapse on either one of these universal branches. When B is the magnetic field driving the SIT at B_{cr} we obtain Fisher's original scaling result [9].

The critical exponents most considered in relation with the SIT are $z\nu = 4/3$ and $z\nu = 8/3$, corresponding to classical and quantum percolation, respectively. The above scaling result, however, is valid only for second-order quantum transitions, characterized by a finite critical exponent ν. BKT transitions are of infinite order and, formally, they have critical exponent $\nu = \infty$, corresponding to a singular behaviour at criticality. For $\eta > 1$ thus, Fisher's result must be modified to take into account the BKT universality class of the two phase transitions between the Bose metal and superconductor and superinsulator, respectively. To do so, we repeat the above analysis with the BKT scaling

$$\xi \propto \mathrm{e}^{\sqrt{\frac{B^*}{|B - B_{\mathrm{cr}}|}}}, \qquad (10.58)$$

where B^* is a constant with dimension of magnetic field. Requiring that temperature scales as the characteristic frequency we obtain

$$T = T_0 \, \mathrm{e}^{-z\sqrt{\frac{B^*}{|B - B_{\mathrm{cr}}|}}}, \qquad (10.59)$$

where T_0 is a non-universal temperature and z a constant. We now follow the same procedure as in the Fisher scaling argument by expressing

$$\ln\left(\frac{T}{T_0}\right) = -z\sqrt{\frac{B^*}{|B - B_{\mathrm{cr}}|}}, \tag{10.60}$$

which leads immediately to the correct scaling variable for a BKT transition,

$$s = |B - B_{\mathrm{cr}}|\left(\ln\left(\frac{T}{T_0}\right)\right)^2. \tag{10.61}$$

Unfortunately, as already mentioned, the quantity T_0 is non-universal and must be fitted.

Three quantum phase transitions in a magnetic-field-driven SIT with an intermediate Bose metal were indeed detected in a NbTiN film [34], as shown in Fig. 10.3. It might be objected that a magnetic field as a driving parameter is not appropriate, since it breaks the protecting symmetry $U(1) \times \mathbb{Z}_2^T$ of the edge modes, opening a gap for them. For the extremely weak fields in the experiment, however, this gap, estimated by the cyclotron frequency, is always smaller than $25\,\mathrm{mK}$, even at the highest applied fields,

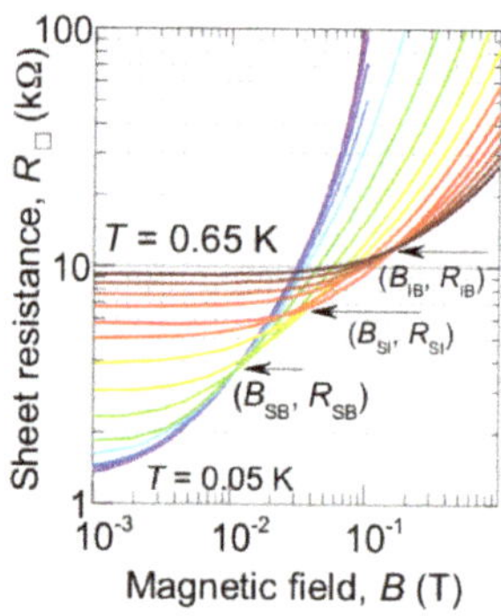

Fig. 10.3. Three crossing points in the resistance dependence on magnetic field at different temperatures in a NbTiN film, indicating the positions of the SIT tricritical point (middle crossing point at the resistance quantum) and the two quantum transitions between the Bose metal and superconductor and Bose metal and superinsulator. From M. C. Diamantini, A. Yu. Mironov, S. V. Postolova, X. Liu, Z. Hao, D. M. Silevitch, Ya. Kopelevich, P. Kim, C. A. Trugenberger and V. M. Vinokur, Bosonic topological intermediate state in the superconductor–insulator transition, *Phys. Lett.* **A384**, 126570 (2020), ©Elsevier (2020).

and much smaller typically. Since the lowest measurement temperatures were 40 mK, the edge modes can be still considered as gapless in the whole experimental range. The three crossing points at $B_{\mathrm{SB}} = 0.011 \pm 0.001$ T, $B_{\mathrm{SI}} = 0.039 \pm 0.001$ T and $B_{\mathrm{IB}} = 0.16 \pm 0.01$ T indicate changes in the resistance behaviour as a function of temperature, i.e. quantum phase transitions. The corresponding resistances are $R_{\mathrm{SB}} = 3.63 \pm 0.01$ kΩ, $R_{\mathrm{SI}} = 6.57 \pm 0.01$ kΩ and $R_{\mathrm{IB}} = 11.9 \pm 0.1$ kΩ, showing that the intermediate resistance R_{SI} is essentially the resistance quantum R_{Q}. The duality relations

$$\frac{B_{\mathrm{SB}}}{B_{\mathrm{SI}}} = \frac{B_{\mathrm{SI}}}{B_{\mathrm{IB}}},$$
$$\frac{R_{\mathrm{SB}}}{R_{\mathrm{SI}}} = \frac{R_{\mathrm{SI}}}{R_{\mathrm{IB}}}, \tag{10.62}$$

predicted by the phase diagram Fig. 8.2 (with g replaced by B here) and Eq. (10.56) are perfectly satisfied within experimental error. This identifies thus $(B_{\mathrm{SI}}, R_{\mathrm{SI}})$ as the SIT quantum tricritical point and the two other couples $(B_{\mathrm{SB}}, R_{\mathrm{SB}})$ and $(B_{\mathrm{IB}}, R_{\mathrm{IB}})$ as the dual superconductor–Bose metal and superinsulator–Bose metal quantum transitions, respectively. Although $\eta > 1$, the critical behaviour around the tricritical point is still clearly observed.

The modified, logarithmic BKT scaling relations at the three quantum transitions are shown in Fig. 10.4. In perfect accord with duality, the best fit parameter T_0 comes out equal, $T_0 = 2$ K for the two dual transitions. It is lower, $T_0 = 1$ K for the SIT tricritical point in the middle. Finally, let us stress that BKT transitions imply, formally, an infinite critical exponent ν. The Harris criterion [112] then fully confirms one more time that disorder becomes irrelevant at the transition fixed points.

To conclude this chapter let us stress its main points. The Bose metal is nothing else than a bosonic topological insulator in which the metallic saturation of the sheet resistance for $T \to 0$ is caused by symmetry-protected ballistic edge modes. The Bose metal experiences two dual quantum BKT transitions to a superconductor and a superinsulator for high and low dimensionless conductances g, respectively. Let us stress, however, that "edges" do not mean necessarily the physical edges of the sample. The quantum phase transitions can be inhomogenous, with separated bubbles of Bose metal forming near the transition points and coalescing to larger and larger bubbles towards the self-dual point $g = 1$. Edge modes can

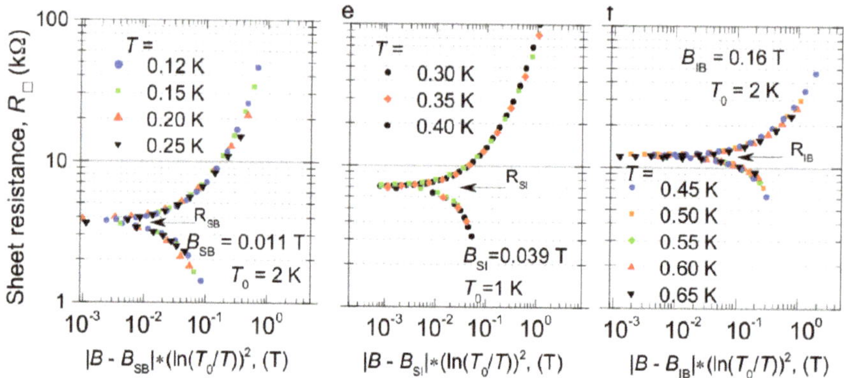

Fig. 10.4. The logarithmic scaling at the three quantum BKT transitions shown in Fig. 10.3. The best fit parameter $T_0 = 2$ K for both the dual transitions between Bose metal and superconductor and Bose metal and superinsulator and $T = 1$ K for the SIT tricritical point in between. From M. C. Diamantini, A. Yu. Mironov, S. V. Postolova, X. Liu, Z. Hao, D. M. Silevitch, Ya. Kopelevich, P. Kim, C. A. Trugenberger and V. M. Vinokur, Bosonic topological intermediate state in the superconductor–insulator transition, *Phys. Lett.* **A384**, 126570 (2020), with permission from the authors.

flow along the edges of these bubbles and tunnel from one to the other if they are sufficiently close. Metallic charge transport can thus flow through a Chalker–Coddington percolation network [113], which is known to be a medium for gapless edge states [114].

Chapter 11

Josephson junction arrays

Until now we have focused on effective field theories, based exclusively on symmetry considerations and topology. The question arises, if there exists a concrete system that realizes the SIT effective field theory at the microscopic level. The answer is affirmative, such a system exists and it is provided by Josephson junction arrays (JJA) [7].

A JJA is a square array with spacing ℓ, comprising superconducting islands with nearest-neighbor Josephson coupling of strength $E_{\rm J}$. Each island has a self-capacitance C_0 and mutual capacitances C to its nearest neighbors. The Hamiltonian for such a JJA is given by [115]

$$H = \sum_{\mathbf{x}} \frac{C_0}{2} V_{\mathbf{x}}^2 + \sum_{<\mathbf{xy}>} \frac{C}{2} \left(V_{\mathbf{y}} - V_{\mathbf{x}}\right)^2 + E_{\rm J} \left(1 - \cos \left(\varphi_{\mathbf{y}} - \varphi_{\mathbf{x}}\right)\right), \quad (11.1)$$

where boldface characters denote the sites of the two-dimensional array, $< \mathbf{xy} >$ indicates nearest neighbours, $V_{\mathbf{x}}$ is the electric potential of the island at $\mathbf{x}$ and $\varphi_{\mathbf{x}}$ the phase of its order parameter. This Hamiltonian can be rewritten as

$$H = \sum_{\mathbf{x}} \frac{1}{2} V_{\mathbf{x}} \left(C_0 - C\nabla^2\right) V_{\mathbf{x}} + \sum_{\mathbf{x},i} E_{\rm J} \left(1 - \cos \left(\Delta_i \varphi_{\mathbf{x}}\right)\right), \quad (11.2)$$

where Δ_i and $\hat{\Delta}_i$ are forward and backward finite differences and ∇^2 is the two-dimensional finite difference Laplacian. The phases $\varphi_{\mathbf{x}}$ are quantum-mechanically conjugated to the charges $\mathcal{E}_{\mathbf{x}}$ on the islands: these are quantized in integer multiples of $2e$ (Cooper pairs), $\mathcal{E}_{\mathbf{x}} = 2eq_{\mathbf{x}}$, $q_{\mathbf{x}} \in \mathbb{Z}$, where e is the electron charge. The Hamiltonian (11.2) can be expressed in terms of charges and phases by noting that the electric potentials $V_{\mathbf{x}}$ are determined by the charges $\mathcal{E}_{\mathbf{x}}$ via a discrete version of Poisson's equation:

$$\left(C_0 - C\nabla^2\right) V_{\mathbf{x}} = \mathcal{E}_{\mathbf{x}}. \tag{11.3}$$

Using this in (11.2) we get

$$H = \sum_{\mathbf{x}} 4E_{\mathrm{C}} \, q_{\mathbf{x}} \frac{1}{C_0/C - \nabla^2} q_{\mathbf{x}} + \sum_{\mathbf{x},i} E_{\mathrm{J}} \left(1 - \cos\left(\Delta_i \varphi_{\mathbf{x}}\right)\right), \tag{11.4}$$

where $E_{\mathrm{C}} \equiv e^2/2C$. The integer charges $q_{\mathbf{x}}$ interact via a two-dimensional Yukawa potential with mass $\sqrt{C_0/C}/\ell$. In the experimentally accessible nearest-neighbors capacitance limit $C \gg C_0$, this implies a two-dimensional Coulomb law at distances smaller than the electrostatic screening length $\Lambda = \ell\sqrt{C/C_0}$. Then the charging energy E_{C} and the Josephson coupling E_{J} are the only two relevant energy scales. These can be traded for one energy parameter $\omega_{\mathrm{P}} = \sqrt{8E_{\mathrm{C}}E_{\mathrm{J}}}$, the Josephson plasma frequency, and one numerical parameter $g = \sqrt{\pi^2 E_{\mathrm{J}}/2E_{\mathrm{C}}}$, the dimensionless conductance. In the following we will consider the physics of JJA at energies much below the plasma frequency, which takes the role of the natural ultraviolet (UV) cutoff in the theory.

The partition function of JJA admits a phase-space path-integral representation [115]

$$Z = \sum_{\{q\}} \int_{-\pi}^{+\pi} \mathcal{D}\varphi \, \exp(-S),$$

$$S = \int_0^\beta dt \left[\sum_{\mathbf{x}} \left(i \, q_{\mathbf{x}} \, \dot{\varphi}_{\mathbf{x}} + 4E_C \, q_{\mathbf{x}} \frac{1}{C_0/C - \nabla^2} q_{\mathbf{x}} \right) \right.$$

$$\left. + \sum_{\mathbf{x},i} E_{\mathrm{J}} \left(1 - \cos\left(\Delta_i \varphi_{\mathbf{x}}\right)\right) \right], \tag{11.5}$$

where $\beta = 1/T$ is the inverse temperature. In (11.5) time has to be considered also as discrete, as generally appropriate when degrees of freedom can change only in integer steps. We introduce thus a discrete time step ℓ_0 and we substitute the time integrals and space sums over a lattice with nodes $\mathbf{x}$ by a sum over space-time lattice nodes x, with x^0 denoting the discrete time direction. Denoting by Δ_0 the (forward) finite time difference we obtain,

$$Z = \sum_{\{q\}} \int_{-\pi}^{+\pi} \mathcal{D}\varphi \, \exp(-S),$$

$$S = \sum_x \left(i \, q_x \Delta_0 \varphi_x + 4\ell_0 E_C \, q_x \frac{1}{C_0/C - \nabla^2} q_x \right)$$

$$+ \sum_{x,i} \ell_0 E_J \left(1 - \cos \left(\Delta_i \varphi_x \right) \right), \tag{11.6}$$

Finally, we use the Villain representation (for a review see [57]) to express the cosine interaction in (11.6) in terms of a set of integer link variables a_i. By introducing real charge currents j_i and assuming that the size of the system is much smaller than the screening length Λ, so that we can safely set $C_0 \to 0$ from now on, we arrive at

$$Z = \sum_{\{a_i\},\{j_0\}} \int \mathcal{D}j_i \int_{-\pi}^{+\pi} \mathcal{D}\varphi \, \exp(-S),$$

$$S = \sum_{x,i} \left(i j_0 \Delta_0 \varphi + i j_i \left(\Delta_i \varphi + 2\pi a_i \right) + 4\ell_0 E_C \, j_0 \frac{1}{-\nabla^2} j_0 + \frac{1}{2\ell_0 E_J} j_i^{\,2} \right),$$

$$\tag{11.7}$$

where we have dropped the underscripts referring to the lattice positions of the variables and we have introduced the notation j_0 for the integer charges.

To proceed further, we stress that Eq. (11.7), derived from what is viewed as the standard JJA Hamiltonian, misses a crucial piece. This missing contribution is a kinetic term proportional to the vortex mass. The omission of this term is common practice when considering overdamped junctions. However, this omission is *a priori* not justified in arrays in which collective effects may lead to a renormalization of the vortex mass, no matter how large its bare value may be. It is known since the very early days of JJA that integrating over charge fluctuations leads anyway to a vortex kinetic term [115]. It is a general principle in field theory that whatever is induced by fluctuations *must* be included at bare level. It is also known that dissipation is substantially reduced when the Coulomb interaction becomes long-range [116], exactly the regime we are interested in. Finally, ballistic motion of these Josephson vortices has indeed been observed experimentally [117]. This calls for the necessity of adding a vortex kinetic term to the action. Such a vortex kinetic term would involve the time derivative of a_i, the integer variable conjugated to the charge currents and would represent phase slips corresponding to one vortex moving from one plaquette to a neighbouring one. When the coefficient of this kinetic term $(\partial_0 a_i)^2$ takes the value $\pi^2/4\ell_0 E_C$ we can introduce a real Lagrange multiplier a_0 and a fictitious electric field $\phi_i = K_{i\mu} a_\mu$ and write the vortex kinetic term and the charge Coulomb interaction compactly as

$$Z = \sum_{\{a_i\},\{j_0\}} \int \mathcal{D}a_0 \mathcal{D}j_i \int_{-\pi}^{+\pi} \mathcal{D}\varphi \, \exp(-S),$$

$$S = \sum_{x,i} \left(i j_0 \left(\Delta_0 \varphi + 2\pi a_0 \right) + i j_i \left(\Delta_i \varphi + 2\pi a_i \right) + \frac{1}{2\ell_0 E_J} j_i^{\,2} + \frac{\pi^2}{4\ell_0 E_C} \phi_i^2 \right).$$

$$(11.8)$$

In this representation, the Coulomb interaction between the charges follows from the Gauss law constraint associated with the Lagrange multiplier a_0. This procedure works only at a particular value of the vortex mass that makes the model self-dual under the interchange of charge and vortex variables, while simultaneously interchanging $\pi^2 E_J \leftrightarrow 2E_C$. As a consequence it was called the self-dual approximation in [6]. We shall adopt this approximation here too.

At this point we note that the charge current j_μ is conserved and, hence, it can be represented as the field strength associated to a second fictitious gauge field b_μ as $j_0 = K_{0i} b_i$, $j_i = K_{i0} b_0 + K_{ij} b_j$, where b_0 is a real variable, while b_i are integers. We then use Poisson's formula,

$$\sum_{n_\mu} f(n_\mu) = \sum_{k_\mu} \int dn_\mu f(n_\mu) \, e^{i2\pi n_\mu k_\mu}, \qquad (11.9)$$

turning a sum over integers $\{n_\mu\}$ into an integral over real variables, to make all components of the gauge fields a_μ and b_μ real, at the price of introducing integer link variables Q_i and M_i,

$$Z = \sum_{\{Q_i\}} \sum_{\{M_i\}} \int \mathcal{D}a_\mu \mathcal{D}b_\mu \int_{-\pi}^{+\pi} \mathcal{D}\varphi \, \exp(-S),$$

$$S = \sum_{x,i} \left[i2\pi \, a_\mu K_{\mu\nu} b_\nu + \frac{1}{2\ell_0 E_J} j_i^{\,2} + \frac{\pi^2}{4\ell_0 E_C} \phi_i^2 + i2\pi a_i Q_i + i2\pi b_i M_i \right.$$

$$\left. + b_i \left(\hat{K}_{i0} \Delta_0 \varphi + \hat{K}_{ij} \Delta_j \varphi \right) + b_0 \hat{K}_{0i} \Delta_i \varphi \right]. \qquad (11.10)$$

Finally, we note that the quantities $(1/2\pi)\hat{K}_{\mu\nu}\Delta_\nu\varphi$ are the circulations of the array phases around the plaquettes orthogonal to the direction μ in 3D Euclidean space-time and are thus quantized as $2\pi \times$ integers. We can thus absorb the quantities $\left(\hat{K}_{i0}\Delta_0\varphi + \hat{K}_{ij}\Delta_j\varphi\right)$ in a redefinition of the integers M_i and define $\hat{K}_{0i}\Delta_i\varphi = 2\pi M_0$. The original integral over the phases φ can then be traded for a sum over the vortex numbers M_0,

$$Z = \sum_{\{Q_i\}} \sum_{\{M_\mu\}} \int \mathcal{D}a_\mu \mathcal{D}b_\mu \, \exp(-S),$$

$$S = \sum_{x,i} \left(i2\pi \, a_\mu K_{\mu\nu} b_\nu + \frac{1}{2\ell_0 E_\mathrm{J}} j_i{}^2 + \frac{\pi^2}{4\ell_0 E_C}\phi_i^2 + i2\pi a_i Q_i + i2\pi b_\mu M_\mu \right).$$

$$(11.11)$$

If we reinstate the canonical dimensions of gauge fields, we rescale them by $1/2\pi$, and we pass to lattice derivatives instead of finite differences, we obtain

$$Z = \sum_{\{Q_i\}} \sum_{\{M_\mu\}} \int \mathcal{D}a_\mu \mathcal{D}b_\mu \, \exp(-S),$$

$$S = \sum_{x,i} \left[i\frac{\ell_0 \ell^2}{2\pi} \, a_\mu k_{\mu\nu} b_\nu + \frac{\ell_0 \ell^2}{8\pi^2 E_\mathrm{J}} j_i{}^2 + \frac{\ell_0 \ell^2}{16 E_C}\phi_i^2 + i\ell a_i Q_i \right.$$
$$\left. + i\left(\ell_0 b_0 M_0 + \ell b_i M_i\right) \right].$$

$$(11.12)$$

This is exactly the $\varepsilon = 1$, $\mu \to \infty$ limit of the effective action (8.5) with the translations

$$ve_v^2 \to 4\pi^2 E_\mathrm{J},$$
$$ve_q^2 \to 8E_C,$$

$$(11.13)$$

so that

$$g \to \sqrt{\frac{\pi^2 E_\mathrm{J}}{2E_C}},$$

$$(11.14)$$

$$\Delta_{\mathrm{top}} = mv^2 = \frac{ve_v e_q}{2\pi} \to \sqrt{8E_C E_\mathrm{J}} = \omega_P.$$

As expected, the dimensionless coupling governing the phase structure of the array is the ratio of the two magnetic and electric energy scales in the problem. The role of the topological energy gap is taken by the plasma frequency, describing small phase oscillations on the superconducting islands of the array. In JJA, the previous effective field theory appears as the exact, microscopic model with the topological energy scale ω_P. For energy scales $\ll \omega_P$ the dynamics of the array is effectively frozen and only magnetic monopole tunneling events on the frequency scale $1/\ell_0 \ll \omega_P$ determine the phase structure. These are events in which vortices appear and disappear on the array, $d_0 M_0 = \pm 1$, accompanied by circular electric field $E^i = \epsilon^{ij} M_j$, as shown in Fig. 9.1. The difference with respect to the effective field theory for films is the duality-breaking absence of the integer variable Q_0. This reflects charge conservation: while a topological charge like the vortex number is not conserved due to quantum tunneling events, a Noether charge originating from a symmetry is strictly conserved. In the effective field theory for films we had to impose this duality-breaking effect by hand, here it appears automatically.

Once established that the microscopic model for JJA coincides with the effective field theory for superconducting films with the translations (11.14) one can use the previously derived results for the quantum phase structure and the string tension of the linearly confining potential due to the monopole instantons in the superinsulating phase,

$$\sigma = \frac{\omega_P}{\ell} \sqrt{\frac{16}{\pi g \ell_0 \omega_P}} \, \mathrm{e}^{-\frac{\pi g}{2\ell_0 \omega_P} G_2(0)}, \qquad (11.15)$$

where $G_2(0)$ is the value of the screened 2D lattice Coulomb potential at coinciding points. The string tension contains two factors. The first, ω_P/ℓ depends solely on the "classical" array parameters, the lattice spacing and the plasma frequency. The second factor depends on the "quantum" characteristics of the array, the coupling g and the ratio between the long tunneling time ℓ_0 and the short phase oscillation period $1/\omega_P$. Since ω_P is the frequency of small phase fluctuations and $\ell_0 \omega_P \gg 1$, the emerging picture of the superinsulating state is that of a frozen phase state in which the dynamics is governed only by quantum tunneling events in which vortices appear and disappear. These tunneling events cause the infinite resistance.

Two of the predicted quantum phases of JJA have been observed. Global superconductivity sets in when all the phases on the islands of the array lock in the same value. The Bose metal has also been detected, most notably in the recent experiment [30], as shown in Fig. 11.1. Note also

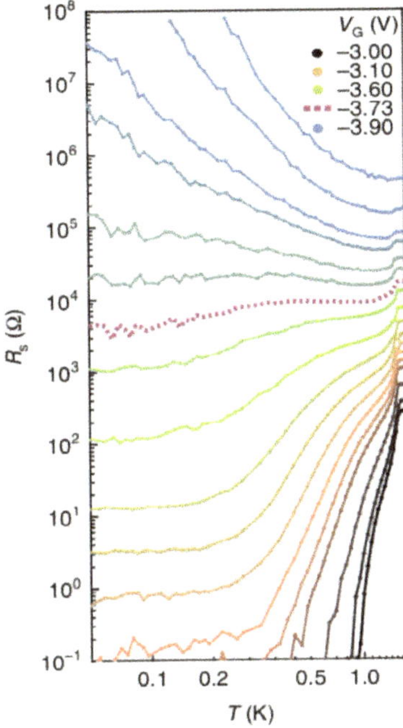

Fig. 11.1. Resistance as a function of temperature in a gated array, with the SIT driven by an external gate voltage. The plot clearly shows three regimes, the superconductor, with resistance dramatically dropping at low temperatures, the Bose metal regime, with a metallic saturation of the resistance at low temperatures, and the insulating regime, with the resistance increasing with decreasing temperature. From C. G. L. Bottcher *et al.*, Superconducting, insulating and anomalous metallic regimes in a gated two-dimensional semiconductor–superconductor array, *Nat. Phys.* **14**, 1138–1144 (2018), with permission by the authors and the publisher. ©Springer (2018)

here the presence of minima at temperatures 0.5–1 K on the insulating side. These temperatures correspond to frequencies of order 10 GHz, which are the typical plasma frequencies of JJA [118]. As in films, these minima indicate the activation of the bulk conduction channel of the topological insulator realized in the Bose metal phase of JJA.

Superinsulation, instead, has not (yet) been observed on JJA. To resolve this puzzle let us estimate the typical size $\ell_{\text{string}} = \sqrt{c\hbar/\sigma}$ of the confining string, where we have reinstated physical units. We take the typical values of experimental JJA [118], $\ell = 100\,\text{nm}$ and $\omega_P = 10\,\text{GHz}$. Then, the first contribution to the string size amounts to $\ell^0_{\text{string}}/\ell = \sqrt{c/\ell\omega_P} \approx 550$. This number is reduced by the second factor in (11.15). However, even for $\ell_0\omega_P = \mathcal{O}(1000)$, we still have $\ell_{\text{string}}/\ell \approx 150$, at the border of the total size, 190, of typical JJA showing the SIT [118]. In field theory parlance, present JJA are too far from the previously derived infrared confining fixed point [46] and, due to asymptotic freedom (for a review see [24]), only the short-distance screened Coulomb forces can be observed. In order to detect superinsulation on JJA one must thus design an array with sufficiently high plasma frequency and with linear size sufficiently large to fit an entire string, presumably in the thousands of lattice spacings. However, one needs also a large screening length $\Lambda/\ell = \sqrt{C/C_0} \gg 1$. As mentioned above, this condition implements experimentally our modeling assumption $C_0 \to 0$ and ensures a proper 2D Coulomb interaction between charges. In JJA, $\Lambda/\ell \simeq$ 20–40 at best. This is insufficient to ensure a 2D Coulomb interaction over the whole array. The major problem hampering the detection of superinsulation in JJA is thus the difficult simultaneous realization of large arrays with an even larger screening length ensuring 2D Coulomb physics over a sufficient portion of the array. Hence, observation of superinsulation in a standard classical JJA at present is hardly possible. A promising platform for highly controllable JJA capable to host superinsulation is offered by proximity arrays that can be driven through the SIT by either a gate voltage [30] or a magnetic field [119].

Chapter 12

Fractional Bose metals

Second-order phase transitions are characterized by scale-invariant correlation functions. The lack of a scale implies that the lattice spacing can be removed at these critical points, giving rise to a continuum field theory. It turns out that critical field theories actually have a larger invariance under the full conformal group. In 3D this has no far reaching consequences. In 2D, however, the conformal group is infinite-dimensional [120]. This implies a huge number of constraints which, ultimately, are sufficient to completely classify all possible critical phenomena in two dimensions [73].

The 2D conformal algebra (Virasoro algebra) has a central extension c [73]. Unitary conformal field theories are characterized by $c \leq 1$, the simplest case for $c = 1$ being the Gaussian model of a single boson ϕ, with action

$$S_G = \frac{1}{2\pi g} \int d^2\mathbf{x}\, (\partial_i \psi)^2. \tag{12.1}$$

Each such theory is a collection of so-called primary fields, representing highest weight states, each with an infinite tower of derived states, called Verma module, obtained by applying the infinite series of lowering operators. The primary fields can be composed among themselves to produce further primary fields, exactly as angular momentum representations can be composed. The corresponding rules are called the fusion rules and the ensemble of primary fields with their fusion rules form the fusion algebra. Primary fields are characterized each by a conformal dimension, ruling the exponent in the power-law decay of its correlation functions. In case of the $c = 1$ Gaussian model the primary fields are given by the so-called vertex operators $\exp(i\sqrt{2}\alpha_i \psi(\mathbf{x}))$ and have conformal dimensions $h_i = \alpha_i^2 g/4$ [73].

Among unitary conformal field theories there is a special series of so-called minimal models, for which the fusion algebra closes on a finite number of primary fields, realizing thus the bootstrap ideal of particle physics: each "particle" is a combination of the others and nothing is fundamental. Minimal models exist for central charge

$$c = 1 - 6\frac{(p - q)^2}{pq},\tag{12.2}$$

where p and q are coprime integers. The conformal dimensions of the finite number of corresponding primary fields are

$$h_{rs} = \frac{(rp - sq)^2 - (p - q)^2}{pq}\tag{12.3}$$

with $1 \leq r \leq q - 1$, $1 \leq s \leq p - 1$, and $|p - q| = 1$ for unitary minimal models. Unitary minimal models can thus be labeled by a single integer p.

Let us now consider equal-time quantum correlation functions in the Bose metal state, in the limit in which we neglect the BKT effects induced by charge or vortex condensation. For simplicity we shall consider only charge density observables, encoded in the field ψ. The same reasoning can of course be adapted to vortex observables. Correlations for generic quantum operators $\mathcal{O}_i\left[\psi\left(\mathbf{x}_i\right)\right]$ follow immediately from the explicit expression (10.37) of the quantum wave functional and are given by

$$\langle\mathcal{O}_1\left[\psi\left(\mathbf{x}_1\right)\right]\ldots\mathcal{O}_m\left[\psi\left(\mathbf{x}_m\right)\right]\rangle = \frac{1}{Z}\int\mathcal{D}\psi\,\mathcal{O}_1\left[\psi\left(\mathbf{x}_1\right)\right]\ldots\mathcal{O}_m\left[\psi\left(\mathbf{x}_m\right)\right]$$

$$\times\,\mathrm{e}^{\frac{-1}{2\pi g}\int d^2\mathbf{x}(\partial_i\psi)^2},$$

$$Z = \int\mathcal{D}\psi\,\mathrm{e}^{\frac{-1}{2\pi g}\int d^2\mathbf{x}(\partial_i\psi)^2}.\tag{12.4}$$

This shows that equal-time quantum correlations in the Bose metal ground state (in the considered approximation) are equivalent to the classical correlations of a $c = 1$ Gaussian model [121]. This is a concrete example of a general class of models in which conformal invariance is found not in the action of a classical model in D-dimensional space but rather in the ground state wave functional of a quantum system in (D+1)-dimensional space-time [122].

One can now apply the well-developed machinery of conformal field theory to generalize the $c = 1$ Bose metal to *fractional Bose metals* with $c < 1$. In this context, the primary fields play the role of excitations above the ground state. The minimal models correspond thus to the only models with a finite set of excitations and should be the most robust possible fractional Bose metals. Conformal minimal models can thus be used to classify fractional Bose metals.

This becomes particularly appealing in view of experimental results on magnetically or electrically frustrated Josephson junction arrays (JJA). We will focus here on magnetic frustration, offset charges can be treated analogously. The crucial parameter for JJA in an external magnetic field B is the frustration $f = \Phi/\Phi_0$ where Φ is the total magnetic flux through the array and $\Phi_0 = \pi/e$ is the fundamental flux quantum. The frustration is thus a parameter defined on the interval $[0,1]$ and describes the number of fundamental magnetic fluxes piercing one plaquette of the JJA modulo 1. What has been observed in experiments [123] is that, for particular values of the frustration f, the array undergoes the usual superconductor-to-insulator transition (SIT), as if the magnetic field would be absent, with the only difference that the position of the SIT is displaced from the self-dual point $1/g = 1$ to lower values. The observed fractions are $f = 1/2, 2/5, 1/3, 1/4, \ldots$ For fully frustrated arrays with $f = 1/2$, the critical parameter $1/g$ for the quantum transition is lowered by a factor approximately 0.7 with respect to the non-frustrated case, with measurements having an error margin of 10% [123]. These results are interpreted as the formation of a vortex lattice commensurate with the underlying junction network, which completely shields the external magnetic field. This mechanism is very reminiscent of the binding of an even number of magnetic fluxes by electrons forming quantum Hall fluids at fractional filling [124], with the only difference that, in the present situation, the magnetic field is totally screened. In the quantum Hall framework it is well known that the emerging composite electrons can be described by dressed vertex operators of conformal field theories [125]. The same applies to the frustrated Bose metal and its SIT tricritical point.

We shall thus assume that the Bose metal in presence of magnetic frustration is described by exactly the same quantum ground state wave functional (10.37) but that a background charge representing the contribution of the shielding vortex lattice has to be included in ground state expectation values. In conformal field theory this is known as the Feigin–Fuchs construction [126], in which the shielding is formally implemented by a charge at infinity:

$$\left\langle e^{i\sqrt{2}\alpha_1\psi(\mathbf{x}_1)}\ldots e^{i\sqrt{2}\alpha_n\psi(\mathbf{x}_n)}\right\rangle_f$$

$$= \lim_{R\to\infty} R^{f^2 g}\left\langle e^{i\sqrt{2}\alpha_1\psi(\mathbf{x}_1)}\ldots e^{i\sqrt{2}\alpha_n\psi(\mathbf{x}_n)}e^{-i\sqrt{2}f\psi(R)}\right\rangle, \qquad (12.5)$$

where the second expectation value is computed with the original ground state wave functional (10.37). When the external magnetic field is completely shielded, the charge at infinity is given exactly by the frustration, so that the only non-vanishing correlations have $\sum_i \alpha_i = f$.

It turns out that the Feigin–Fuchs representation corresponds to a shift of the central charge and of the conformal dimensions of primary fields [126],

$$c = 1 \rightarrow c = 1 - 6\frac{f^2 g}{4},$$

$$h_i = \frac{\alpha_i^2 g}{4} \rightarrow h_i = \frac{\alpha_i(\alpha_i - f)g}{4}. \tag{12.6}$$

It is natural to posit that the observed frustrated Bose metals corrrespond to the unitary conformal minimal models, since these are the most stable ground states with a finite number of available excitations. In the Feigin–Fuchs construction, the minimal models are obtained for $f = 2(q - p)/q$ and $1/g = p/q$ for p and q coprime integers and $|p - q| = 1$. Setting $p = m$ and $q = m + 1$ we obtain the following series of models,

$$f = \frac{2}{m + 1},$$

$$\frac{1}{g} = \frac{m}{m + 1}, \tag{12.7}$$

$$m \geq 3.$$

The first few predicted frustrations are

$$f = \frac{1}{2}, \frac{2}{5}, \frac{1}{3}, \frac{2}{7}, \frac{1}{4}, \dots \tag{12.8}$$

Apart from 2/7, these are exactly the observed frustrations where a SIT analogous to the $f = 0$ one is observed. The fraction 2/7 corresponds probably to a higher state in the hierarchy which is not sharp enough for present measurements to detect it. Note also that the predicted critical coupling of the fully frustrated model with $f = 1/2$ is $1/g = 3/4 = 0.75$. This is well within the 10% error margin of the experimental result 0.7. In conclusion, like in the case of the quantum Hall fluids, the powerful methods of conformal field theory are enough to predict the series of fractional Bose metals in presence of frustration. Of course, the BKT effects due to charge and vortex condensation in these frustrated Bose metals have still to be added in.

Chapter 13

Symmetry classification
of superinsulator excitations

In the preceding chapter we have learned about the power of generic symmetry considerations to establish the quantum numbers of admitted excitations and, thereby, to provide an exhaustive classification of possible Bose metal ground states in presence of frustration. This is an example of a *dynamical symmetry*. A dynamical symmetry is an organizational principle for the classical configuration space and the quantum mechanical Hilbert space based on a fundamental symmetry of the problem [127]. The admitted states in the model are the representations of the symmetry algebra and are characterized by quantum numbers corresponding to the values of the Cartan subalgebra generators in these representations. The Hamiltonian is expanded in the Casimir operators of the symmetry algebra so that each representation has the same energy for all its states [127]. Excitations can be combined by the representation combination rules, called fusion rules.

One very well-known example of dynamical symmetry is the flavour SU(3) group of the quark model (for a review see [20]). Once established that this is the correct dynamical symmetry of the strong interaction, one can immediately classify the possible elementary particles as irreducible representations of the unitary group SU(3). The fundamental representations 3 and $\bar{3}$ are the quarks and antiquarks themselves. The representations obtained by composing 3 and $\bar{3}$ are the mesons, a singlet and an octet. Representations made of three quarks are, for example, the baryon decuplet and so on. A further example of dynamical symmetry is the conformal symmetry characterizing critical phenomena in 2D [73]. As we have seen in the preceding chapter, where we used this symmetry to classify frustrated Bose metals, this is an infinite symmetry. On one side this makes the representation theory more complex, on the other side, however, it admits minimal

115

models on which the fusion algebra closes, representing thus particularly stable models.

The simplest excitations in superinsulators are electrically neutral, static couples of ± 2 charges bound together by a string of electric flux and corresponding to the scalar mesons of the strong interaction. In this case, the next-higher excitations are rotating couples of quark–antiquark pairs, corresponding to vector mesons with spin. The relation between the square mass of these particles and spin is linear: it is called a Regge trajectory (for a review see [4]). Since both the flavour and colour groups are absent in superinsulators we do not expect the richness of the hadronic spectrum. We do, however, expect the existence of excitations with spin.

To find the dynamical symmetry for superinsulators we start from the observation that all ground states with a gap for density waves are incompressible. In the limit of infinite gap, the dynamical symmetry algebra of a 2D incompressible configuration is the algebra w_∞ of area-preserving diffeomorphism [128]. Correspondingly, the Hilbert space of gapless excitations of a 2D quantum incompressible ground state can be organized in representations of the algebra of quantum area-preserving diffeomorphisms. For a chiral state this is an extension of the conformal algebra, called $W_{1+\infty}$. This dynamical symmetry was used to classify all possible chiral quantum incompressible fluid plateaus of the quantum Hall effect [74,75]. For a non-chiral state it is the direct product of right and left copies of the same algebra, $W_{1+\infty} \otimes \overline{W}_{1+\infty}$. Each algebra $W_{1+\infty}$ contains a U(1) subalgebra. The vector diagonal $U_V(1)$ (the "sum" of the two chiral U(1)) is the electromagnetic algebra and classifies charges, the axial diagonal $U_A(1)$ (the "difference" of the two chiral U(1)) is the dual algebra classifying vortices.

Both superconductors and superinsulators are ground states with a gap for density waves (charge-density waves are admitted in d-wave high-T_c superconductors with gapless nodal points, this is however a different order competing with superconductivity and the gapless nodal points don't affect incompressibility when superconductivity dominates). In both, however, the electromagnetic vector $U_V(1)$ cannot be part of the dynamical symmetry algebra, albeit for a different reason. In 2D superconductors, charge is not a good quantum number because $U_V(1)$ is realized in the "Kosterlitz–Thouless mode". Note that this is *not* spontaneous symmetry breaking, which is impossible in 2D (apart when the temperature is strictly zero). There is no order parameter related to a Higgs field, the superfluid density is expressed exclusively in terms of the power-law correlations of Josephson vortices with no inner core [47] (see Sec. 14.2). In superinsulators, instead, only electrically neutral excitations are admitted, so that $U_V(1)$ is also absent from the dynamical symmetry algebra. As a consequence, the correct dynamical symmetry for both 2D superconductors and superinsulators is

$$W = \frac{W_{1+\infty} \otimes \overline{W}_{1+\infty}}{U_V(1)}. \tag{13.1}$$

This dynamical symmetry algebra was originally proposed in [129] for superconductors alone. As we now show, one of the two possible universality classes found there corresponds actually to superinsulators.

Area-preserving diffeomorphisms are canonical transformations of a two-dimensional phase space. Let us endow the plane with dimensionless coordinates z and $\bar{z}$, normalized by the typical length scale of the problem, in the case of superconductors the coherence length, and with a Poisson bracket

$$\{f, g\} = \pm i(\partial f \bar{\partial} g - \bar{\partial} f \partial g). \tag{13.2}$$

One can then describe area-preserving diffeomorphisms $\delta z = \{\mathcal{L}, z\}$ and $\delta \bar{z} = \{\mathcal{L}, \bar{z}\}$ in terms of generating functions $\mathcal{L}(z, \bar{z})$. The basis of generators $\mathcal{L}_{n,m} = z^n \bar{z}^m$ satisfies the classical w_∞ algebra [128]

$$\{\mathcal{L}_{n,m}, \mathcal{L}_{k,l}\} = \mp i(mk - nl)\mathcal{L}_{n+k-1,m+l-1}. \tag{13.3}$$

Note that the operators $\mathcal{L}_{nm}$ with $n > 0$ and $m > 0$ form two closed subalgebras related by complex conjugation. These are the two chiral sectors of the classical w_∞ algebra. Generators with both n and m negative are descendants that can be obtained as products of primary generators in the two fundamental chiral sectors. Finally, generators with $n = 0$ and $m = 0$ form two Abelian subalgebras. To describe a parity-invariant incompressible system we can thus use the upper/lower signs in Eqs. (13.2) and (13.3) for the chiral sectors $n \geq 0$ and $m \geq 0$.

The quantum version of this infinite-dimensional algebra is obtained by the usual substitution of Poisson brackets with quantum commutators: $i\{\,,\} \to [\,,]$. Let us denote the quantum version of $\mathcal{L}_{i-n,i}$ by V_n^i and let us first discuss one single chiral sector of the quantum algebra by restricting to positive values of i. This gives the algebra W_∞,

$$[V_n^i, V_m^j] = (jn - im)\, V_{n+m}^{i+j-1} + q(i, j, n, m)\, V_{n+m}^{i+j-3} + \cdots$$
$$+ \delta^{ij}\delta_{n+m,0}\, c\, d(i, n), \tag{13.4}$$

where the structure constants q and d are polynomials of their arguments and the dots denote a finite number of similar terms involving the operators $V_{n+m}^{i+j-1-2k}$. The first term on the r.h.s. of (13.4) is the classical term (13.3).

The remaining terms are quantum operator corrections, with the exception of the last c-number term, which represents a quantum anomaly with central charge c. All the quantum operator corrections are uniquely determined by the closure of the algebra; only the central charge c is a free parameter. If we admit also the value $i = 0$ we obtain the full algebra $W_{1+\infty}$, including the Abelian subalgebra quantum generators V_n^0. The quantization of the full classical algebra w_∞, involving generators in the two chiral sectors, the two Abelian subalgebras, and all their products, is then obtained as the direct product of two copies $W_{1+\infty}$ and $\overline{W}_{1+\infty}$ of opposite chirality.

The generators V_n^i are characterized by an integer conformal (scaling) dimension $h = i + 1 \geq 1$ and an angular momentum mode index n, $-\infty < n < +\infty$. The operators V_n^0 satisfy the Abelian Kac–Moody algebra (for a review see [73]) $\widehat{U}(1)$, which is the quantum extension of the usual U(1) by a c-number central charge, while the operators V_n^1 are the generators of conformal transformations, satisfying the Virasoro algebra (for a review see [73]),

$$[V_n^0, V_m^0] = n\, c\, \delta_{n+m,0},$$

$$[V_n^1, V_m^0] = -m\, V_{n+m}^0, \tag{13.5}$$

$$[V_n^1, V_m^1] = (n - m)V_{n+m}^1 + \frac{c}{12}n(n^2 - 1)\delta_{n+m,0}.$$

The operators V_n^0 and V_n^1 are the charge and angular-momentum modes in the chiral sector under consideration.

A quantum model with $W_{1+\infty}$ dynamical symmetry is defined as a Hilbert space constructed as a set of irreducible, unitary, highest-weight representations of $W_{1+\infty}$, which is closed under the fusion rules for making composite states. The irreducible, unitary, quasi-finite, highest-weight representations of $W_{1+\infty}$ have been completely classified in [130]. They exist only if the central charge is a positive integer, $c = m \in \mathbb{Z}_+$. They are characterized by an m-dimensional weight vector $\vec{r}$ with real elements and are built on top of a highest weight state $|\vec{r}\rangle_W$ which satisfies

$$V_n^i|\vec{r}\rangle_W = 0, \qquad \forall\, n > 0,\ i \geq 0, \tag{13.6}$$

and is an eigenstate of the V_0^i,

$$V_0^i|\vec{r}\rangle_W = \sum_{n=1}^{m} m^i(r_n)\, |\vec{r}\rangle_W, \tag{13.7}$$

where $m^i(r)$ are i-th order polynomials of a weight component. In particular, the charge $q = V_0^0$ and scaling dimension $h = V_0^1$ are given by

$$q = \sum_{n=1}^{m} m^0(r_n) = r_1 + \cdots + r_m,$$

$$h = \sum_{n=1}^{m} m^1(r_n) = \frac{1}{2}\left[(r_1)^2 + \cdots + (r_m)^2\right].$$

(13.8)

Note that scaling dimension and angular momentum coincide in a chiral theory.

The incompressible quantum ground state is the special highest-weight state $|\Omega\rangle_W$ satisfying

$$V_n^i|\Omega\rangle_W = 0, \quad \forall\, n > 0\ ,\ i \geq 0,$$

$$V_0^i|\Omega\rangle_W = 0, \quad i \geq 0.$$

(13.9)

Particle–hole excitations are obtained by applying generators with negative mode index to $|\Omega\rangle_W$. Due to incompressibility, these gapless excitations are edge excitations. Other highest-weight representations in the $W_{1+\infty}$ model are identified with further possible bulk incompressible states, each with its tower of gapless edge excitations. These representations encode the possible bulk quasi-particle excitations and are characterized by an infinite set of quantum numbers (13.7). Since the representations that we consider are quasi-finite, the number $d(n)$ of independent edge excitations at total angular momentum level n, encoded in the character of the representation, is finite.

There exist two types of irreducible, unitary, quasi-finite highest-weight representations of $W_{1+\infty}$ [130]. For *generic* representations the weight vector $\vec{r}$ is such that $(r_i - r_j) \notin \mathbb{Z}$, $\forall\, i \neq j$. These representations are equivalent to the corresponding $\widehat{U}(1)^{\otimes m}$ representations with the same weight. *Degenerate* representations, instead, have $(r_i - r_j) \in \mathbb{Z}$ for some $i \neq j$. The weight components $\{r_i\}$ of the degenerate representations can be grouped and ordered in congruence classes modulo $\mathbb{Z}$ [130]. A representation with two classes is the tensor product of two one-class representations. Therefore, the one-class degenerate representations are actually the basic building blocks for all degenerate representations. These one-class representations have the weight vectors

$$\vec{r} = \{r_1, \ldots, r_m\} = \{s + n_1, \ldots, s + n_m\}, \quad s \in \mathbb{R}, \quad n_1 \geq \cdots \geq n_m \in \mathbb{Z}.$$

(13.10)

They are in one-to-one relation with the representations of U(1)⊗SU(m). In fact, by an orthogonal transformation it is possible to introduce a new basis [75] $\vec{q} = \{q, \mathbf{\Lambda}\}$ for the $W_{1+\infty}$ weights, so that

$$q = \frac{1}{\sqrt{m}} \left(r_1 + r_2 + \cdots + r_m \right),$$

$$\Lambda_a = \sum_{i=1}^{m} u_a^{(i)} \, r_i, \qquad a = 1, \ldots, m-1, \tag{13.11}$$

where $\mathbf{u}^{(i)}$ are the weight vectors of the defining SU(m) representation [127]. Each such representation embodies therefore one charged excitation and $(m-1)$ neutral excitations with a hidden SU(m) symmetry.

In the $\{q, \mathbf{\Lambda}\}$ basis the fusion rules for making composite $W_{1+\infty}$ representations take a particularly simple form:

$$\vec{q} \bullet \vec{p} = \vec{p} + \vec{q} \quad \mod \quad \left\{ \begin{pmatrix} 0 \\ \alpha^{(1)} \end{pmatrix}, \cdots, \begin{pmatrix} 0 \\ \alpha^{(m-1)} \end{pmatrix} \right\},$$

$$\alpha^{(a)} = \mathbf{u}^{(a)} - \mathbf{u}^{(a+1)}, \qquad a = 1, \ldots, m-1. \tag{13.12}$$

This means that charge is additive, while the neutral excitations combine according to the SU(m) fusion rules, i.e. the SU(m) weights add up modulo an integer combination of the simple roots $\alpha^{(a)}$.

The one-class degenerate $W_{1+\infty}$ representations are in one-to-one equivalence with those of the $\widehat{U}(1) \otimes \mathcal{W}_m$ minimal models, where $\mathcal{W}_m$ is the Fateev–Lykyanov–Zamolodchikov algebra [131] in the limit $c_{\mathcal{W}_m} \to m-1$. The minimality is reflected in a lower number $d(n)$ of independent edge excitations at level n than for generic representations because of additional relations among the states, leading to null vectors which have to be projected out in order to maintain irreducibility [73]. This is the origin of the reducibility of the $\widehat{U}(1)^{\otimes m}$ representations with respect to the $W_{1+\infty}$ algebra. $W_{1+\infty}$ *minimal models*, defined as $W_{1+\infty}$ models made only of one-class degenerate representations, are thus particularly robust quantum ground states, with a minimal set of excitations with respect to generic models. In the chiral case, the $W_{1+\infty}$ minimal models are in one-to-one correspondence with the observed hierarchy of quantum Hall states [75]. In the following we shall focus on non-chiral W minimal models as a classification of all possible 2D superconductors and superinsulators.

The chiral $W_{1+\infty}$ minimal models were constructed in [75]. They are built from weights belonging to lattices $\mathcal{L}$ which are closed under the fusion rules (13.12). In the original $\vec{r}$ basis, these lattices $\vec{r} = \sum_{i=1}^{m} n_i \, \vec{v}_i$ are generated by the basis vectors

$$(\vec{v}_i)_j = \delta_{ij} + r C_{ij}, \tag{13.13}$$

where $r \in \mathbb{R}$, $C_{ij} = 1$, $\forall i, j = 1, \ldots, m$ and the integer excitation labels n_i must obey the constraints $n_1 \geq n_2 \geq \cdots \geq n_m$ in order to avoid double counting. The real number r originates from the free parameter s of the one-class degenerate representations in (13.10). Using the basis (13.11) and the fact that $\sum_{i=1}^{m} \mathbf{u}^{(i)} = 0$, it is easy to recognize that the excitations having no neutral component are of the form $n_i = n$, $\forall i = 1, \ldots, m$. This defines the $\widehat{U}(1)$ axis of the lattice. The charge unit is thus given by the lattice vector $\sum_{i=1}^{m} \vec{v}_i$ and the charge of a generic excitation is the linear projection of its lattice vector on this unit charge vector. The charge and scaling dimension of the excitation with label $\{\mathbf{n}\}$ are thus given by

$$
\begin{aligned}
q &= \mathbf{t}^T \cdot M \cdot \mathbf{n}, \\
h &= \frac{1}{2} \, \mathbf{n}^T \cdot M \cdot \mathbf{n},
\end{aligned}
\tag{13.14}
$$

where $\mathbf{t} = (1, \ldots, 1)$ and the matrix M is the metric of the lattice $\mathcal{L}$:

$$M_{ij} = \vec{v}_i \cdot \vec{v}_j = \delta_{ij} + \lambda \, C_{ij}, \quad \lambda = mr^2 + 2r. \tag{13.15}$$

As a consequence, the charge unit of the theory is determined by the parameter r as $q_{\text{unit}} = m \, (1 + mr)^2$. Note that the spectrum of the theory contains also fractionally charged excitations.

The conjugate algebra of opposite chirality is spanned by the complex conjugate generators $\bar{z}^{i-n} z^i$ with $i \geq 0$ and can be obtained using the quantum commutator $[z, \bar{z}] = +1$. The generators $\overline{V}_n^i = \bar{z}^{i-n} z^i$ satisfy thus exactly the same algebra (13.4) as the original operators V_n^i. We shall call the algebra spanned by these generators $\overline{W}_{1+\infty}$. A weight vector $(r_1, \ldots, r_n)$ of $W_{1+\infty}$ is also a weight vector of $\overline{W}_{1+\infty}$ and the weights $\bar{m}$, $\bar{q}$, and $\bar{h}$ are given by the same polynomial expressions (13.7) and (13.8). In this sector of opposite chirality, however, the angular momentum coincides with the negative $-\bar{h}$ of the scaling dimension. A minimal model of $W_{1+\infty} \otimes \overline{W}_{1+\infty}$ at level $c = m$ is thus spanned by a lattice $\mathcal{L} \otimes \mathcal{L}$ with $\mathcal{L}$ defined in (13.13). Its excitations have integer labels $\{\mathbf{n}, \bar{\mathbf{n}}\}$ and charge, vortex number, spin, and scaling quantum numbers

$$Q = q + \bar{q}, \quad \Phi = q - \bar{q},$$
$$S = h - \bar{h}, \quad H = h + \bar{h}. \tag{13.16}$$

To obtain the minimal models of the relevant coset dynamical symmetry W in (13.1) we must finally divide out the diagonal, vector $\widehat{U}(1)_V$ Kac–Moody algebra identified with electric charge. This can be accomplished by restricting to lattices $\mathcal{L}$ for which the total charge vanishes identically, $Q = 0$. There are then no charge excitations in the spectrum and the generators $V^0_{-n} + \overline{V}^0_{-n}$ can be consistently eliminated.

Using Eqs. (13.16) and (13.14) it is easy to rewrite the condition $Q = 0$ as $Q = (1 + mr)^2 \sum_i (n_i + \bar{n}_i) = 0$. There are two solutions to this equation. The first is a generic solution and consists of restricting to excitations with quantum numbers $\{\mathbf{n}, \bar{\mathbf{n}} = \mathbf{n}^R\}$, where the reflected excitation R is defined by $n_1^R = -n_m, \ldots, n_m^R = -n_1$, so that both constraints $Q = 0$ and $\bar{n}_1 \geq \bar{n}_2 \geq \cdots \geq \bar{n}_m$ are satisfied. In this case, the full vector $(W_{1+\infty})_V$ is broken and the dynamical symmetry reduces to

$$W = (W_{1+\infty})_A. \tag{13.17}$$

The excitation spectrum reduces to bosonic vortices with fluxes $\Phi = k\Phi_0$ which are integer multiples of the flux quantum $\Phi_0 = 2(1 + mr)^2$. The gapped quantum ground states with broken $\widehat{U}_V(1)$ are superconductors. These are the only solution at level $c = 1$, where the full dynamical symmetry reduces essentially to $\widehat{U}_A(1)$ and labels the vortex excitations.

The most intriguing situation occurs, instead, when the parameter r assumes the value $r = -1/m$. In this case the chiral lattice $\mathcal{L}$ is degenerate and consists entirely of neutral excitations: no additional conditions have to be imposed in order to divide out the diagonal subgroup $\widehat{U}_V(1)$. Actually, this has the consequence that the vortex number Φ also vanishes identically, so that both $\widehat{U}_V(1)$ and $\widehat{U}_A(1)$ are effectively divided out and the residual dynamical symmetry becomes

$$W = W_m \otimes \overline{W}_m, \tag{13.18}$$

with W_m the Fateev–Lykyanov–Zamolodchikov algebra [131] in the limit $c_{W_m} \to m - 1$. The interpretation of these gapped ground states is obvious: charge vanishes because it is confined, vortices are absent from the spectrum because they are condensed. This is a superinsulator and the residual neutral states represent its possible excited states.

As we have seen in preceding chapters, the simplest excitation in a superinsulator is a spinless, neutral bound state of two ± 2 charges. This excitation does not carry any W quantum number that can distinguish it from the ground state. Only on symmetry grounds, however, we have now established that there exist classes of superinsulators labeled by an integer $c = m$ with also neutral spinon excitations in their spectrum

$$S = h - \bar{h},$$

$$h = \frac{1}{2} \sum_i n_i^2 - \frac{1}{m} \left(\sum_i n_i \right)^2, \tag{13.19}$$

$$\bar{h} = \frac{1}{2} \sum_i \bar{n}_i^2 - \frac{1}{m} \left(\sum_i \bar{n}_i \right)^2.$$

A particularly suitable basis for these $(m - 1)$ elementary excitations is given by the vectors $(\mathbf{n}^{(a)}, \mathbf{0})$ with $\mathbf{n}^{(a)}$ defined by $n_i^{(a)} = 1$, $\forall i \leq a$ and $n_i^{(a)} = 0$, $\forall i > a$. The corresponding weight vectors in the basis (13.11) become then the $\mathrm{SU}(m)$ fundamental weights,

$$\mathbf{\Lambda} = \mathbf{\Lambda}^{(a)} = \sum_{i=1}^{a} \mathbf{u}^{(i)}. \tag{13.20}$$

Each elementary excitation is thus associated with the highest weight of an $\mathrm{SU}(m)$ fundamental representation and has spin

$$S^{(a)} = \frac{a}{2} \left(1 - \frac{a}{m} \right), \tag{13.21}$$

the lowest possible value being

$$S_{\min} = \frac{m - 1}{2m}. \tag{13.22}$$

The conjugate excitations of opposite chirality have the same structure with spins of opposite sign. The spinons are thus anyons [79], the simplest example being the semions (half-fermions) at level $c = 2$. At each level $c = m$ they carry also an $\mathrm{SU}(m)$ isospin quantum number: their fractional statistics is therefore non-Abelian. Note, however, that the $\mathrm{SU}(m)$ symmetry of these excitations is different from the usual symmetry of the quark model of strong interactions (for a review see [20]). Indeed, spinons do not come in full SU(m) multiplets; rather, only the highest-weight states are present. These, however, combine according to the usual SU(m) fusion rules, which explains the non-Abelian character of their monodromies.

Chapter 14

Three dimensions: A gauge theory for vortices

14.1. The 3D SIT

Electric interactions follow a Coulomb law, i.e. an inverse linear potential in 3D and a logarithmic potential in 2D. Magnetic interactions between vortices are always logarithmic. One way to realize the superconductor-to-insulator transition (SIT), the one discussed so far, is to squeeze a superconductor into a thin film so that electric Coulomb interactions become 2D, i.e. logarithmic and thus comparable with vortex interactions. There is another, dual one, however: making superconducting vortices loose and *open*, with *magnetic monopoles* at their endpoints. Magnetic monopoles have the same 3D inverse linear interactions as electric charges and the long, loose, open vortices between monopole–antimonopole pairs play the role of unobservable Dirac strings, as far as the long-distance/low-energy physics is concerned. Instead of changing the electric interaction to make it comparable to the magnetic one, we "modify" the magnetic interaction to make it comparable with the electric one. This leads to the 3D SIT and 3D superinsulators, first discussed in [6, 16].

How would a material that realizes this mechanism look like? Consider, e.g. a layered, inhomogeneous system of superconducting granules, i.e. bubbles of Cooper pair condensate coupled by tunneling links. Each such bubble is characterized by an independent phase of the superconducting order parameter. The circulation of these phases on groups of adjacent bubbles in the same layer can form 2D vortex configurations, so-called pancake vortices [132]. These pancakes typically align along the z-axis to form "composite" 3D vortex lines resembling those found in isotropic superconductors. These stacks of pancake vortices then end in magnetic monopole–antimonopole pairs on the surfaces of the sample. When, however, the horizontal mobility of ballistic pancake vortices is taken into account, the stacks can break. These breaking points then appear as magnetic monopoles in the interior of the sample. The prototype of this type of material is a 3D Josephson junction array (JJA) consisting of a stack of 2D JJA coupled by Josephson links also in the vertical direction [6]. Of course, a layered material with pancake vortices is just an example to explain the principle of how open vortices can form. Such open vortices can appear in any granular superconductor. The inhomogenous superconductors with emergent granularity [35] which realize this mechanism have been predicted in [36] and subsequently experimentally confirmed in [37]. It is important to stress that the open vortices we are considering here are not Abrikosov vortices within a Bose condensate but, rather, Josephson vortices [5], forming in between granules of condensate. As such they do not have a normal core and neither do the magnetic monopoles at their endpoints. These are thus free to Bose condense and to move dissipationlessly.

The relevant degrees of freedom in 3D granular superconductors are single Cooper pairs, which can tunnel from one granule to the next, leaving behind a Cooper hole, and Josephson vortices, which, to begin with, we will consider as closed loops. These are the excitations that can acquire topological Aharonov–Bohm–Casher phases when one is encircling the other. Is a local representation generalizing Wilczek's Chern–Simons construction (8.1) [55] possible also in 3D? The answer is yes, although point-like charges couple, as usual, to gauge fields a_μ, while vortices couple to Kalb–Ramond antisymmetric tensor fields $b_{\mu\nu}$ [133], also called gauge fields of the second kind,

$$\mathcal{L} = \frac{1}{4\pi} b_{\mu\nu} \epsilon^{\mu\nu\alpha\beta} \partial_\alpha a_\beta + a_\mu j^\mu + \frac{1}{2} b_{\mu\nu} m^{\mu\nu}, \qquad (14.1)$$

where j^μ and $m^{\mu\nu}$ are the charge and vortex currents, respectively. Since vortices are now one-dimensional objects, their conserved current is an antisymmetric tensor, containing generically 6 degrees of freedom, two orthogonal directions for each direction in which the vortex can point. The action (14.1) is called the BF action (for a review see [134]) and it is the 3D generalization of a mixed Chern–Simons term. It was introduced as a field theory for a condensed matter system in [6]. The BF model is topological, since it is metric-independent. It is invariant under the usual gauge transformations $a_\mu \to a_\mu + \partial_\mu \xi$ but also under gauge transformations of the second kind,

$$b_{\mu\nu} \to b_{\mu\nu} + \partial_\mu \lambda_\nu - \partial_\nu \lambda_\mu. \tag{14.2}$$

Finally, if a_μ is a vector field and $b_{\mu\nu}$ a pseudotensor field, the model is also invariant under the parity ($\mathcal{P}$, with group $\mathbb{Z}_2^{\mathrm{P}}$) and time-reversal ($\mathcal{T}$, with group $\mathbb{Z}_2^{\mathrm{T}}$) symmetries. In general, the BF action for a model defined on a compact space with non-trivial topology has a ground state degeneracy [135] reflecting the topology, exactly as the mixed Chern–Simons term in 2D. However, when the coefficient of the BF term is $1/4\pi$, as in (14.1), the ground state is unique [135]. In this case, (14.1) is the effective field theory for the simplest 3D bosonic topological insulator [136].

The field strength associated to a_μ is, as usual, $f_{\mu\nu} = \partial_\mu a_\nu - \partial_\nu a_\mu$. In 3D, the dual field strength is a 2-tensor,

$$\tilde{f}^{\mu\nu} = \frac{1}{2}\epsilon^{\mu\nu\alpha\beta} f_{\alpha\beta} = \epsilon^{\mu\nu\alpha\beta}\partial_\alpha a_\beta. \tag{14.3}$$

Since $b_{\mu\nu}$ is itself a 2-tensor, instead, its field strength is a 3-tensor

$$h_{\mu\nu\alpha} = \partial_\mu b_{\nu\alpha} + \partial_\nu b_{\alpha\mu} + \partial_\alpha b_{\mu\nu} . \tag{14.4}$$

It is easy to check that this is invariant under gauge transformations of the second kind (14.2). The dual field strength tensor is a vector,

$$h^\mu = \frac{1}{6}\epsilon^{\mu\nu\alpha\beta} h_{\nu\alpha\beta} = \frac{1}{2}\epsilon^{\mu\nu\alpha\beta}\partial_\nu b_{\alpha\beta}. \tag{14.5}$$

When using the BF term to model the emergent behaviour of condensed matter systems, one identifies the fluctuation components of the topologically conserved charge current j_{fluc}^μ and vortex current $m_{\mathrm{fluc}}^{\mu\nu}$ as

$$\begin{aligned}
j_{\mathrm{fluc}}^\mu &= \frac{1}{2\pi}h^\mu = \frac{1}{4\pi}\epsilon^{\mu\nu\alpha\beta}\partial_\nu b_{\alpha\beta}, \\
m_{\mathrm{fluc}}^{\mu\nu} &= \frac{1}{2\pi}\tilde{f}^{\mu\nu} = \frac{1}{2\pi}\epsilon^{\mu\nu\alpha\beta}\partial_\alpha a_\beta.
\end{aligned} \tag{14.6}$$

For Cooper pairs, charges are measured in integer units of $2e$ and vortices in integer units of $2\pi/2e = \pi/e$.

The field strengths $f_{\mu\nu}$ and $h_{\mu\nu\alpha}$ can be used to add dynamics to the purely topological BF term,

$$\mathcal{L} = \frac{1}{12\Lambda^2}h_{\mu\nu\alpha}h^{\mu\nu\alpha} + \frac{1}{4\pi}b_{\mu\nu}\epsilon^{\mu\nu\alpha\beta}\partial_\alpha a_\beta - \frac{1}{4f^2}f_{\mu\nu}f^{\mu\nu}, \tag{14.7}$$

where f is a dimensionless coupling and Λ has canonical dimension [1/length]. Repeating the same derivation as in 2D one obtains the equation of motion

$$\left(\Box + m^2\right) \tilde{f}^{\mu\nu} = 0,$$

$$\left(\Box + m^2\right) h^\mu = 0, \qquad (14.8)$$

$$m = \frac{f\Lambda}{2\pi}.$$

The BF term is thus the 3D equivalent of the 2D Chern–Simons topological mass term [137]. In this gauge-invariant model, the "densities" b_{0i} never appear with time derivatives in the action. They are thus not dynamical variables but, rather, Lagrange multipliers enforcing three constraints analogous to the Gauss law in usual gauge theories. The remaining three components b_{ij} can be transformed by time-independent gauge transformations of the second kind, with a vector function λ_i as in (14.2). Of the three components of this vector, however, only two are real gauge functions, since λ_i itself is invariant under usual gauge transformations $\lambda_i \to \lambda_i + \partial_i \xi$. Therefore the three spatial components b_{ij} contain one single degree of freedom. Gauge invariance of the second kind implies a substantial reduction of the degrees of freedom.

Let us consider now the Euclidean-space action corresponding to (14.7), coupled to point sources representing charges and Josephson vortices and constituting, as in 2D, the topological excitations of the model,

$$S = \int d^4 x \left(\frac{1}{12\Lambda^2} h_{\mu\nu\alpha} h_{\mu\nu\alpha} + \frac{i}{4\pi} b_{\mu\nu} \epsilon^{\mu\nu\alpha\beta} \partial_\alpha a_\beta + \frac{1}{4f^2} f_{\mu\nu} f_{\mu\nu} \right.$$

$$\left. + i a_\mu j_\mu + \frac{i}{2} b_{\mu\nu} m_{\mu\nu} \right), \qquad (14.9)$$

where, as in 2D, all zero components of vectors are rescaled by the light velocity $v = 1/\sqrt{\varepsilon\mu}$ in the medium so that $x^0 = vt$. By integrating out the two gauge fields with the techniques explained in earlier chapters we obtain

$$S_{m\to\infty} \left(j_\mu, m_{\mu\nu}\right) = \lim_{m\to\infty} \left(-i \ln \int \mathcal{D}a_\mu \mathcal{D}b_{\mu\nu} e^{-S(a_\mu, b_{\mu\nu}, j_\mu, m_{\mu\nu})} \right)$$

$$= \lim_{m\to\infty} \int d^4 x \; \pi m^2 j_\mu \frac{\epsilon_{\mu\nu\alpha\beta} \partial_\nu}{\nabla^2 \left(m^2 - \nabla^2\right)} m_{\alpha\beta}. \qquad (14.10)$$

By representing the conserved charge current as $j_\mu = (1/2)\epsilon_{\mu\nu\alpha\beta}\partial_\nu n_{\alpha\beta}$, with $\partial_\mu n_{\mu\nu} = \partial_\nu n_{\mu\nu} = 0$ to maintain the correct number of degrees of freedom, we obtain

$$S_{m \to \infty}\left(j_\mu, m_{\mu\nu}\right) = 2\pi \int d^4 x \; \frac{1}{2} n_{\mu\nu} m_{\mu\nu}, \tag{14.11}$$

which, for quantized point-like sources, reduces to an integer multiple of 2π representing the intersection number of a transverse curve and a surface in 4 Euclidean dimensions. The BF model is indeed the generalization of Wilczek's local representation of the topological interactions of point particles and vortices in 3D. The spatial scale $1/vm$ represents the gauge size of charges and Josephson vortices, which is the UV cutoff in the theory. On scales much larger than this, both appear as point-like objects.

To obtain the effective field theory of the SIT in 3D it remains only to formulate (14.9) on a 4D lattice with spatial spacing ℓ and Euclidean time spacing $\ell_0 = \ell/v$, ℓ representing the typical size of the superconducting granules. To this end, however, we must first introduce the lattice BF term. If d_μ, $\hat{d}_\mu$, S_μ, and $\hat{S}_\mu$ denote forward and backward lattice derivatives and shifts, as in (4.1), then the forward and backward lattice BF terms are defined by the three-index operators [6]

$$\begin{aligned}
k_{\mu\nu\rho} &\equiv S_\mu \epsilon_{\mu\alpha\nu\rho} d_\alpha, \\
\hat{k}_{\mu\nu\rho} &\equiv \epsilon_{\mu\nu\alpha\rho} \hat{d}_\alpha \hat{S}_\rho,
\end{aligned} \tag{14.12}$$

where no summation over equal indices μ and ρ is implied. The two lattice BF operators are interchanged (no minus sign) upon summation by parts on the lattice and are gauge invariant,

$$\begin{aligned}
k_{\mu\nu\rho} d_\nu &= k_{\mu\nu\rho} d_\rho = \hat{d}_\mu k_{\mu\nu\rho} = 0, \\
\hat{k}_{\mu\nu\rho} d_\rho &= \hat{d}_\mu \hat{k}_{\mu\nu\rho} = \hat{d}_\nu \hat{k}_{\mu\nu\rho} = 0.
\end{aligned} \tag{14.13}$$

They also satisfy the identities

$$\hat{k}_{\mu\nu\rho} k_{\rho\lambda\omega} = -\left(\delta_{\mu\lambda}\delta_{\nu\omega} - \delta_{\mu\omega}\delta_{\nu\lambda}\right) \nabla^2 + \left(\delta_{\mu\lambda} d_\nu \hat{d}_\omega - \delta_{\nu\lambda} d_\mu \hat{d}_\omega\right)$$

$$+ \left(\delta_{\nu\omega} d_\mu \hat{d}_\lambda - \delta_{\mu\omega} d_\nu \hat{d}_\lambda\right),$$

$$\hat{k}_{\mu\nu\rho} k_{\rho\nu\omega} = k_{\mu\nu\rho} \hat{k}_{\rho\nu\omega} = 2\left(\delta_{\mu\omega} \nabla^2 - d_\mu \hat{d}_\omega\right), \tag{14.14}$$

where $\nabla^2 = \hat{d}_\mu d_\mu$ is the lattice Laplacian. As before, we use the notation Δ_μ and $\hat{\Delta}_\mu$ for the forward and backward finite difference operators and the capital letters $K_{\mu\nu\alpha}$ and $\hat{K}_{\mu\nu\alpha}$ for the corresponding lattice BF terms.

At this point we can formulate the full effective field theory of the 3D SIT,

$$S = \sum_x \left(\frac{\ell^4}{4f^2} f_{\mu\nu} f_{\mu\nu} + i \frac{\ell^4}{4\pi} a_\mu k_{\mu\alpha\beta} b_{\alpha\beta} + \frac{\ell^4}{12\Lambda^2} h_{\mu\nu\alpha} h_{\mu\nu\alpha} \right.$$
$$\left. + i\ell a_\mu Q_\mu + i\ell^2 \frac{1}{2} b_{\mu\nu} M_{\mu\nu} \right). \tag{14.15}$$

The dimensionless parameter $f = O(e)$ encodes the effective Coulomb interaction strength in the material, Λ is the magnetic scale, $\Lambda = O(1/\lambda_\mathrm{L})$, where λ_L is the London penetration depth of the superconducting granules. The topological mass is given by $m = f\Lambda/2\pi v$. Finally, Q_μ and $M_{\mu\nu}$ are integer link and plaquette variables that describe the "Euclidean world-lines" of point charges and "Euclidean world-surfaces" of vortices on the lattice.

Until now we have considered closed vortex loops. Now let us introduce magnetic monopoles at the endpoints of open vortices. When open vortices are present, the gauge symmetry of the second kind (14.2) is broken and the longitudinal components of the tensor gauge field $b_{\mu\nu}$ become usual vector gauge fields for the magnetic monopoles. These induce for the monopoles the same type of Coulomb interaction experienced by charges and this is the basis for the 3D SIT. However, this Coulomb interaction is subdominant with respect to the linear tension created by the vortices between a monopole–antimonopole pair and we will thus neglect it for the determination of the phase structure in the following. Concretely, when inverting the Kalb–Ramond kernel, we will consider only the transverse components of the vortices and neglect their endpoints.

We follow exactly the same procedure as in 2D, i.e. we integrate out the emergent gauge fields to obtain an effective action for the charges and vortices alone,

$$S_\mathrm{top} = \sum_x \left(\frac{f^2}{2\ell^2} Q_\mu \frac{\delta_{\mu\nu}}{(mv)^2 - \nabla^2} Q_\nu + \frac{\Lambda^2}{8} M_{\mu\nu} \frac{\delta_{\mu\alpha}\delta_{\nu\beta} - \delta_{\mu\beta}\delta_{\nu\alpha}}{(mv)^2 - \nabla^2} M_{\alpha\beta} \right), \tag{14.16}$$

where we have left out, for simplicity of presentation, the topological linking term which does not contribute at large distances anyway. The topological interactions, however, also induce screened, short-range interactions, which we approximate by a constant and which represent the bare mass and tension of point charges and line vortices. World-lines and world-surfaces are thus assigned "energies" proportional to their length N and area A (measured in numbers of links and plaquettes),

$$S_{\mathrm{N}} = 2\pi(mv\ell)G \, \frac{f}{\Lambda\ell} \, Q^2 N,$$

$$S_{\mathrm{A}} = 2\pi(mv\ell)G \, \frac{\Lambda\ell}{f} \, M^2 A,$$
(14.17)

where $G = G(mv\ell)$ is the diagonal element of the lattice kernel $G(x-y)$ representing the inverse of the operator $\ell^2\left((mv)^2 - \nabla^2\right)$, and Q and M are the integer quantum numbers carried by the two kinds of excitations. However, the "entropy" of link strings and plaquette surfaces is also proportional to their length and area [138], $\mu_{\mathrm{N}} N$ and $\mu_{\mathrm{A}} A$. Both coefficients μ are non-universal: $\mu_{\mathrm{N}} \simeq \ln(7)$ since, at each step, the non-backtracking string can choose among 7 possible directions on how to continue, while μ_{A} does not have such a simple interpretation but can be estimated numerically. This gives for both types of excitations a "free energy" proportional to their dimension and with coefficients that can be positive or negative depending on the parameters of the theory. The total free energy is

$$F = 2\pi(mv\ell)G \left[\left(\frac{f}{\Lambda\ell} \, Q^2 - \frac{1}{\eta_{\mathrm{Q}}} \right) N + \left(\frac{\Lambda\ell}{f} \, M^2 - \frac{1}{\eta_{\mathrm{M}}} \right) A \right],$$

where we have defined

$$\eta_{\mathrm{Q}} = \frac{2\pi(mv\ell)G}{\mu_{\mathrm{N}}}, \qquad \eta_{\mathrm{M}} = \frac{2\pi(mv\ell)G}{\mu_{\mathrm{A}}}.$$
(14.18)

The coefficients of N and A in this "free energy" represent actually the quantum renormalized mass and tension of the two excitations. If they are positive, the self-energy dominates and large string/surface configurations are suppressed in the partition function. In this regime Cooper pairs/vortices are gapped excitations, suppressed at zero temperature by their mass/tension. The large vortex tension implies that vortices are short and magnetic monopoles correspondingly confined, as shown in panel (a) of Fig. 14.1. If the renormalized mass/tension, are negative instead, the "entropy" dominates and large configurations are favoured. The phase in which long world-lines of Cooper pairs dominate the Euclidean partition function is a charge Bose condensate, i.e. a superconductor. The phase in which large world-surfaces of vortices dominate, instead, is a magnetic monopole condensate, i.e. a 3D superinsulator. In this phase the vortices between monopole–antimonopole pairs become long and loose, as shown in panel (b) of Fig. 14.1. On distance scales $\gg 1/vm$ these long vortices

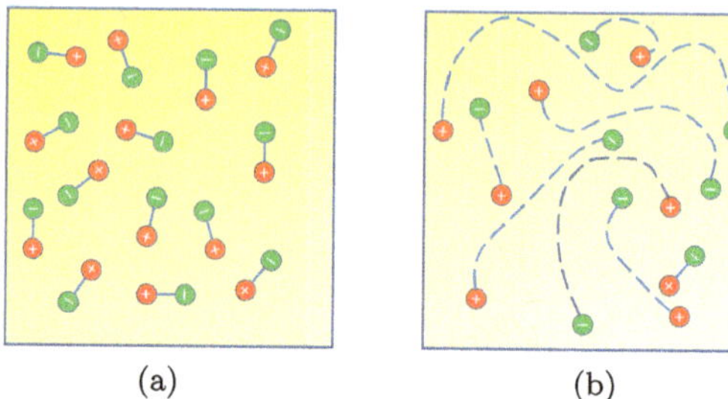

Fig. 14.1. Magnetic monopole states at low temperatures. (a) Monopoles are confined into massive dipoles by the tension of vortices connecting them. (b) As the tension vanishes, vortices become long and loose and magnetic monopoles at their endpoints condense. From M. C. Diamantini, C. A. Trugenberger, V. M. Vinokur, Quantum magnetic monopole condensate, *Nature Comm. Phys.* **4**, 25 (2021). Creative Commons Attribution 4.0.

become unobservable Dirac strings. The world-lines of magnetic monopoles at the endpoints of the loose vortices simultaneously also become long, which means that the deconfined magnetic monopoles Bose condense.

As in 2D, the combined energy–entropy balance equations are best viewed as defining the interior of an ellipse on a 2D integer lattice of electric and magnetic quantum numbers,

$$\frac{Q^2}{r_Q^2} + \frac{M^2}{r_M^2} < 1, \tag{14.19}$$

where the semiaxes are given by

$$r_Q^2 = \frac{\ell\Lambda}{f} \frac{1}{\eta_Q} = \frac{\ell\Lambda}{f} \sqrt{\frac{\mu_N}{\mu_A}} \frac{1}{\eta},$$

$$r_M^2 = \frac{f}{\ell\Lambda} \frac{1}{\eta_M} = \frac{f}{\ell\Lambda} \sqrt{\frac{\mu_A}{\mu_N}} \frac{1}{\eta}, \tag{14.20}$$

with

$$\eta = \sqrt{\eta_Q \eta_M} = 2\pi(mv\ell)G/\sqrt{\mu_N \mu_A}. \tag{14.21}$$

Contrary to 2D, configurations with $Q \neq 0$ and $M \neq 0$ must be excluded since the two types of excitations are different, only pairs $\{0, M\}$ or $\{Q, 0\}$ have to be considered. The phase diagram is found by establishing which integer charges lie within the ellipse when the semiaxes are varied.

This gives

$$\eta < 1 \rightarrow \begin{cases} g < 1, \text{charge Bose condensate,} \\ g > 1, \text{monopole Bose condensate,} \end{cases}$$

$$\eta > 1 \rightarrow \begin{cases} g < \frac{1}{\eta}, \text{charge Bose condensate,} \\ \frac{1}{\eta} < g < \eta, \text{bosonic insulator,} \\ g > \eta, \text{monopole Bose condensate,} \end{cases} \tag{14.22}$$

which is the same quantum phase diagram as in 2D, represented pictorially in Fig. 8.2, but with the tuning parameter g modified to

$$g = \frac{f}{\ell \Lambda} \sqrt{\frac{\mu_{\rm N}}{\mu_{\rm A}}}. \tag{14.23}$$

In 3D, duality is broken by the non-universal ratio of link-to-plaquette entropies, reflecting the different dimensionality of the condensing objects. In this respect it is also important to point out that the condensation of surfaces is more involved than that of lines. When the action is a pure area term the partition function is typically dominated by pathological, "fingered" surfaces, akin to branched polymers, or crumpled surfaces [138]. However, when the action contains a non-local, massive kernel like in (14.16), the condensing surfaces are indeed well-behaved smooth objects of Hausdorff dimension $d_H = 2$ [139], as we will show below.

As in 2D, to distinguish the various phases around the SIT we compute the electromagnetic response by coupling an external electromagnetic gauge potential A_μ to the charge current $j_\mu = (1/2\pi)h_\mu$,

$$S \rightarrow S - i \sum_x \ell^4 A_\mu j_\mu = S - i \sum_x \ell^4 \frac{1}{4\pi} A_\mu k_{\mu\alpha\beta} b_{\alpha\beta}, \tag{14.24}$$

and integrating over the fictitious gauge fields a_μ and $b_{\mu\nu}$. This requires no new computation since, by a summation by parts, the above coupling amounts only to a shift

$$M_{\mu\nu} \rightarrow M_{\mu\nu} - \frac{1}{2\pi} \ell^2 \hat{k}_{\mu\nu\alpha} A_\alpha, \tag{14.25}$$

in (14.16). At long distances $\gg 1/mv\ell$ we obtain thus

$$S_{\rm top}\left(Q_\mu, M_{\mu\nu}, A_\mu\right) = \sum_x \left[\frac{1}{8f^2} \left(\tilde{\mathcal{F}}_{\mu\nu} - 2\pi M_{\mu\nu} \right)^2 - iQ_\mu A_\mu + \frac{2\pi^2}{(\ell\Lambda)^2} Q_\mu^2 \right], \tag{14.26}$$

where we have introduced again dimensionless gauge fields $\mathcal{A}_\mu = \ell A_\mu$ and $\mathcal{F}_{\mu\nu} = \ell^2 F_{\mu\nu}$, as in the corresponding 2D formula.

14.2. Topologically ordered (Higgsless) superconductivity

It is easy to recognize from this expression that a sum over charges Q_μ leads to a mass term for the gauge field A_μ, i.e. to the London equations, as in 2D. As expected, the charge Bose condensate is a superconductor. As we now show, however, this is not a standard superconductor, described by a long-distance Ginzburg–Landau effective field theory. To see this let us go back to our effective gauge field theory (14.1). In the bosonic topological insulator phase, to be discussed below, charges and vortices are short-lived, massive excitations and the ground state is characterized by $j^\mu = 0$ and $m^{\mu\nu} = 0$. The superconductor is realized when charges condense. In this case we have $m^{\mu\nu} = 0$ and j^μ can be promoted to a fluctuating field describing the condensate. Since the current is transverse we set $j^\mu = (1/4\pi)\epsilon^{\mu\nu\alpha\beta}\partial_\nu c_{\alpha\beta}$. The effective field theory then becomes

$$\mathcal{L} = \frac{1}{4\pi} a_\mu \epsilon^{\mu\nu\alpha\beta} \partial_\nu \left(b_{\alpha\beta} + c_{\alpha\beta} \right) + \dots \, , \tag{14.27}$$

where the dots denote terms of higher order in derivatives. Recalling that the IR-dominant BF term is the gauge-invariant mass term in 3D, we see that we are left with one massless Goldstone mode $(b_{\mu\nu} - c_{\mu\nu})$ and one massive mode $(b_{\mu\nu} + c_{\mu\nu})$. The massless Goldstone mode gives mass to the photon when electromagnetic fields are coupled. Contrary to the standard Anderson–Higgs mechanism, however, there is no scalar Higgs field. Rather, associated with this superconductivity mechanism there is a massive vector field whose three degrees of freedom are encoded in a_μ (two of them) and $(b_{\mu\nu} + c_{\mu\nu})$ (the remaining one) [105]. The absence of the scalar Higgs field is reflected in the Josephson-like character of the vortices forming in the granular superconducting medium.

In this type of superconductivity, the opening of a gauge gap is not due to spontaneous symmetry breaking and the Anderson–Higgs mechanism but to a topological mechanism: correspondingly it was called "topologically ordered superconductivity" [140] or "Higgsless" superconductivity [105]. The coherence length associated with the pairing gap is also not related to a normal core representing a defect in a Higgs field but is, instead, the scale of the self-organized granules in the material. The character of vortices in the two types of superconductivities is thus different: Abrikosov vortices with a

Higgs core for spontaneous symmetry breaking, Josephson vortices for topologically ordered superconductivity. Correspondingly, topologically ordered superconductors have a self-organized granular structure and undergo a quantum phase transition to topological insulators or superinsulators. Granular, topologically ordered superconductivity [105,140] with an accompanying massive vector excitation is the only possible superconductivity in 2D, it is a genuine alternative in 3D, where it has recently been detected [37].

14.3. 3D superinsulators

The electromagnetic response $S_{\text{eff}}(A_\mu)$ in the magnetic monopole condensate is given by

$$\mathrm{e}^{-S_{\text{eff}}(A_\mu)} = \sum_{M_{\mu\nu}} \mathrm{e}^{-\frac{1}{8f^2}\sum_{x,\mu,\nu}\left(\tilde{\mathcal{F}}_{\mu\nu}-2\pi M_{\mu\nu}\right)^2}. \tag{14.28}$$

The monopole condensation for strong f turns the real electromagnetic field into a compact variable, defined on the interval $[-\pi,+\pi]$ and the electromagnetic response is given again by Polyakov's compact QED action [18,19], this time in 3D. As in 2D, this changes drastically the Coulomb interaction. To show this we follow the same route as in Sec. 5.3, by considering two external probe charges $\pm q_{\text{ext}}$ and computing the expectation value for the corresponding Wilson loop operator $W(C)$, where C is a closed loop, now in 4D Euclidean space-time,

$$\langle W(C)\rangle$$
$$= \frac{1}{Z_{A_\mu,M_{\mu\nu}}} \sum_{\{M_{\mu\nu}\}} \int_{-\pi}^{+\pi} \mathcal{D}A_\mu \; \mathrm{e}^{-\frac{1}{8f^2}\sum_{x,\mu,\nu}\left(\tilde{\mathcal{F}}_{\mu\nu}-2\pi M_{\mu\nu}\right)^2} \mathrm{e}^{iq_{\text{ext}}\sum_C l_\mu A_\mu}, \tag{14.29}$$

where $l_\mu = 1$ on the links forming the closed loop C and $l_\mu = 0$ everywhere else. When the loop C is restricted to the plane formed by the Euclidean time and one of the space coordinates, $\langle W(C)\rangle$ measures the interaction energy between charges $\pm q_{\text{ext}}$. A perimeter law indicates a short-range potential, while an area-law is tantamount to a linear interaction between them [18,19].

For small enough values of the coupling f, the action is peaked around the values $\tilde{\mathcal{F}}_{\mu\nu} = 2\pi M_{\mu\nu}$, allowing for the saddle-point approximation to compute the Wilson loop. Using the lattice Stoke's theorem $l_\mu = \ell\hat{k}_{\mu\alpha\beta}S_{\alpha\beta}$, one rewrites Eq. (14.29) as

$$\langle W(C) \rangle = \frac{1}{Z_{A_\mu, M_{\mu\nu}}} \sum_{\{M_{\mu\nu}\}} \int_{-\pi}^{+\pi} \mathcal{D}A_\mu \; e^{-\frac{1}{8f^2} \sum_x \left(\tilde{\mathcal{F}}_{\mu\nu} - 2\pi M_{\mu\nu} \right)^2}$$

$$\times \; e^{i \frac{q_{\text{ext}}}{2} \sum_S S_{\mu\nu} \left(\tilde{\mathcal{F}}_{\mu\nu} - 2\pi M_{\mu\nu} \right)}, \tag{14.30}$$

where the quantities $S_{\mu\nu}$ are unit surface elements perpendicular (in 4D) to the plaquettes forming the surface S encircled by the loop C and vanish on all other plaquettes. We have also multiplied the Wilson loop operator by 1 in the form $\exp(-i\pi q_{\text{ext}} \sum_x S_{\mu\nu} M_{\mu\nu})$. Following Polyakov [18,19], we decompose $M_{\mu\nu}$ into transverse and longitudinal components,

$$M_{\mu\nu} = M^{\text{T}}{}_{\mu\nu} + M^{\text{L}}{}_{\mu\nu},$$
$$M^{\text{T}}{}_{\mu\nu} = \ell \hat{k}_{\mu\nu\alpha} n_\alpha + \ell \hat{k}_{\mu\nu\alpha} \xi_\alpha, \tag{14.31}$$
$$M^{\text{L}}{}_{\mu\nu} = \Delta_\mu \lambda_\nu - \Delta_\nu \lambda_\mu,$$

where $\{n_\mu\}$ are integers and we adopt the gauge choice $\Delta_\mu \lambda_\mu = 0$, so that $\nabla^2 \lambda_\mu = \hat{\Delta}_\nu \Delta_\nu \lambda_\mu = m_\mu$, with $m_\mu \in \mathbb{Z}$ describing the world-lines of the magnetic monopoles on the lattice. The set of 6 integers $\{M_{\mu\nu}\}$ has thus been traded for 3 integers $\{n_\mu\}$ and 3 integers $\{m_\mu\}$ representing the magnetic monopoles. The former are then used to shift the integration domain for the gauge field A_μ to $[-\infty, +\infty]$. The real variables $\{\xi_\mu\}$ can then also be absorbed into the gauge field. The integral over the now non-compact gauge field A_μ gives the Gaussian fluctuations around the saddle points m_μ. Gaussian fluctuations contribute the usual Coulomb potential $1/|\mathbf{x}|$ in 3D. We shall henceforth focus only on the magnetic monopoles,

$$\langle W(C) \rangle = \frac{1}{Z_{m_\mu}} \sum_{\{m_\mu\}} e^{-\frac{\pi^2}{2f^2} \sum_{x,\mu} m_\mu \frac{1}{-\nabla^2} m_\mu} \; e^{i2\pi q_{\text{ext}} \sum_S m_\mu \frac{1}{-\nabla^2} \hat{\Delta}_\nu S_{\nu\mu}}.$$

$$\tag{14.32}$$

Following [141] we introduce a dual gauge field χ_μ with field strength $g_{\mu\nu} = \Delta_\mu \chi_\nu - \Delta_\nu \chi_\mu$ and we rewrite (14.32) as

$$\langle W(C) \rangle = \frac{1}{Z_{m_\mu, \chi_\mu}} \int \mathcal{D}\chi_\mu \; e^{-\frac{f^2}{2\pi^2} \sum_{x,\mu,\nu} g_{\mu\nu}^2}$$

$$\times \sum_N \frac{z^N}{N!} \sum_{x_1, \ldots, x_N} \sum_{m_\mu^1, \ldots, m_\mu^n = \pm 1} e^{i \sum_{x,\mu} m_\mu (\chi_\mu + q_{\text{ext}} \eta_\mu)}, \tag{14.33}$$

where the angle $\eta_\mu = 2\pi \hat{\Delta}_\nu S_{\nu\mu}/(-\nabla^2)$ represents a dipole sheet on the Wilson surface S and the monopole fugacity z is determined by the self-interaction as

$$z = \mathrm{e}^{-\frac{\pi^2}{2f^2}G(0)}, \tag{14.34}$$

with $G(0)$ being the inverse of the Laplacian at coinciding arguments. We also used the dilute monopole approximation, valid at sufficiently large f, in which one takes into account only single monopoles $m_\mu = \pm 1$. The sum can now be explicitly performed [141], with the result

$$\langle W(C)\rangle = \frac{1}{Z_{\chi_\mu}} \int \mathcal{D}\chi_\mu \, \mathrm{e}^{-\frac{f^2}{2\pi^2}\sum_{x,\mu,\nu} g_{\mu\nu}^2 + \frac{4\pi^2}{f^2}z(1-\cos(\chi_\mu + q_{\mathrm{ext}}\eta_\mu))}. \tag{14.35}$$

By shifting the gauge field χ_μ by $-q_{\mathrm{ext}}\eta_\mu$ and introducing $M^2 = (\pi^2/f^2)z$, we can rewrite this as

$$\langle W(C)\rangle = \frac{1}{Z_{\chi_\mu}} \int \mathcal{D}\chi_\mu \, \mathrm{e}^{-\frac{2f^2}{\pi^2}\sum_{x,\mu,\nu} \frac{1}{4}g'_{\mu\nu}{}^2 + M^2(1-\cos(\chi_\mu))}, \tag{14.36}$$

where $g'_{\mu\nu} = g_{\mu\nu}(\chi_\mu - q_{\mathrm{ext}}\eta_\mu)$.

For sufficiently large f, this integral is dominated by the classical solution to the equation of motion

$$\hat{\Delta}_\mu g^{\mathrm{cl}}_{\mu\nu} = -2\pi q_{\mathrm{ext}}\hat{\Delta}_\mu S_{\mu\nu} + M^2 \sin \chi^{\mathrm{cl}}_\nu. \tag{14.37}$$

Let us assume that the Wilson loop lies in the (0-3) plane formed by the Euclidean time direction 0 and the z-axis. In this case, there are non-trivial solutions only for the 1- and 2-components of the gauge field, while $\chi^{\mathrm{cl}}_3 = 0$. With the ansatz $\chi^{\mathrm{cl}}_1 = \chi^{\mathrm{cl}}_1(x_2)$, $\chi^{\mathrm{cl}}_2 = \chi^{\mathrm{cl}}_2(x_1)$, we are left with two one-dimensional equations in the region far from the boundaries of the Wilson surface S,

$$\hat{\Delta}_1\hat{\Delta}_1\chi^{\mathrm{cl}}_2 = -2\pi q_{\mathrm{ext}}\hat{\Delta}_1 S_{12} + M^2 \sin \chi^{\mathrm{cl}}_2,$$
$$\hat{\Delta}_2\Delta_2\chi^{\mathrm{cl}}_1 = -2\pi q_{\mathrm{ext}}\hat{\Delta}_2 S_{21} + M^2 \sin \chi^{\mathrm{cl}}_1. \tag{14.38}$$

Following [18, 19], we solve these equations in the continuum limit,

$$\partial_1\partial_1\chi^{\mathrm{cl}}_2 = 2\pi q_{\mathrm{ext}}\delta'(x_1) + M^2 \sin \chi^{\mathrm{cl}}_2,$$
$$\partial_2\partial_2\chi^{\mathrm{cl}}_1 = 2\pi q_{\mathrm{ext}}\delta'(x_2) + M^2 \sin \chi^{\mathrm{cl}}_1. \tag{14.39}$$

For $q_{\text{ext}} = 1$ (corresponding to Cooper pairs in our case), the classical solutions with the boundary conditions $\chi^{\text{cl}}_{1,2} \to 0$ for $|x_{1,2}| \to \infty$ are

$$\chi^{\text{cl}}_1 = \text{sign}(x_2)\ 4\ \arctan \text{e}^{-M|x_2|},$$
$$\chi^{\text{cl}}_2 = \text{sign}(x_1)\ 4\ \arctan \text{e}^{-M|x_1|}. \tag{14.40}$$

Inserting this back in (14.36) we get an area law

$$\langle W(C) \rangle = \text{e}^{-\sigma A}, \tag{14.41}$$

where A is the area of the surface S enclosed by the loop C and the string tension is given by

$$\sigma = \frac{32f}{\pi\sqrt{\varepsilon\mu}} \frac{1}{\ell^2} \exp\left(-\frac{\pi G(0)}{8f^2}\right), \tag{14.42}$$

where $G(0) = 0.155$ is the value of the 4D lattice Coulomb potential at coinciding points. The monopole condensate, thus, generates a string binding together charges and preventing charge transport in systems of a sufficient size. A magnetic monopole condensate is a 3D superinsulator, characterized by an infinite resistance at finite temperatures [6, 16, 36]. The critical value of the effective Coulomb interaction strength for the transition to the superinsulating phase is $f_{\text{crit}} = O(\ell/\lambda_L)$.

14.4. 3D bosonic insulators

The phase with no condensates has the electromagnetic response

$$S_{\text{eff}}(A_\mu) = \frac{1}{8f^2} \sum_{x,\mu,\nu} \ell^4 F^2_{\mu\nu} \tag{14.43}$$

of an insulator. The long-distance effective field theory reduces to

$$\mathcal{L} = \frac{1}{4\pi} a_\mu \epsilon^{\mu\nu\alpha\beta} \partial_\nu b_{\alpha\beta}, \tag{14.44}$$

where we have restored continuous Minkowski space-time notation. As in 2D, this is a purely topological action: essentially all bulk dynamics is suppressed by a large mass gap of topological origin, leaving low-energy degrees of freedom only on the surface. This state is the simplest 3D bosonic topological insulator [136]. Actually, it is a so-called trivial bosonic topological insulator, to be distinguished from a strong bosonic topological insulator [136], as we will discuss more in depth below.

Contrary to 2D, where the edge is 1D, this state is not a Bose metal. Since there is no boson–fermion equivalence on the 2D boundary, the massless surface degrees of freedom remain bosonic and describe thus a surface superfluid. To see this we follow exactly the same procedure as originally introduced by Wen for the edge states of the incompressible quantum Hall fluids [82]. We choose the Weyl gauge $a_0 = 0$ and $b_{0i} = 0$ in which the non-dynamical Lagrange multipliers are eliminated from the action. The effective action (14.44) remains non-invariant on the boundary under the residual static gauge transformations $a_i \to a_i + \partial_i \xi$ and $b_{ij} \to b_{ij} + \partial_i \lambda_j - \partial_j \lambda_i$. To restore full gauge invariance we have to add new dynamical fields ξ and λ_i on the boundary ∂M with a compensating Lagrangian

$$\mathcal{L}_{\partial M} = b\dot{\xi} - Mv_s^2 \left(\partial_i \xi\right)^2 - \frac{1}{2M}b^2, \tag{14.45}$$

where $b = (1/2\pi)\epsilon^{ij}\partial_i \lambda_j$, M is a non-universal mass scale and v_s is the propagation speed of the surface modes.

The phase ξ of the gauge field is the dynamical surface variable, the charge b the corresponding conjugated momentum. The surface Hamiltonian density is

$$\mathcal{H}_{\partial M} = \frac{1}{2M}p^2 + Mv_s^2(\partial_i \xi)^2. \tag{14.46}$$

The Hamiltonian equations of motion

$$\left(\partial_t^2 - v_s^2 \sum_{i=1}^{2} \partial_i^2\right) \xi = 0 \tag{14.47}$$

show that this is a massless Goldstone boson associated to long-wave fluctuations of the phase. The surface is thus a superfluid state with spontaneously broken U(1) symmetry, while the discrete $\mathbb{Z}_2^T$ time-reversal symmetry is preserved [136]. Does this mean that Bose metals cannot exist in 3D? The answer is no, however they are realized differently than in 2D, as we show below.

14.5. Finite-temperature phase transitions

Magnetic monopoles have Coulomb interactions, as can be easily recognized from (14.32). It could thus be expected that the finite-temperature confinement–deconfinement transition they induce is in the universality class of the 3D XY model [57, 59]. For pure compact QED this is indeed the case [142]. In the present model, however, compact QED is induced as the long-distance electromagnetic response in a theory containing both monopoles and charges. As a consequence, what matters for the transition is not the Coulomb interaction of the monopoles but, actually, the dominant tension of the vortices joining them which, as a consequence of the topological interactions of the two types of matter, become physical, observable objects on scales $O(\lambda_L)$, see (14.16). One way to analyze the transition is to consider the finite-temperature behaviour of the charge-confining strings (see Chapter 17) [139], including both the lattice scale ℓ and the topological scale $1/vm = O(\lambda_L)$. The result [143] is that the finite-temperature confinement–deconfinement transition is in the Vogel–Fulcher–Tamman class [144], a quasi-2D behaviour in which the correlation length at the transition diverges according to the law

$$\xi_{\text{corr}} \propto e^{\frac{\xi}{|T-T_{\text{cr}}|}}, \tag{14.48}$$

to be compared with the 2D XY model behaviour (5.20), the two differing only by the power in the exponent. This critical behaviour has been detected in InO films [145], which, having typical thicknesses $d = O(40)\ \xi_{\text{coherence}}$ are very thick compared to films of other materials, with $d = O(1)\ \xi_{\text{coherence}}$.

What about the superconducting transition? Here too, the topological charge–vortex interaction screens the otherwise Coulomb interaction on the scale $1/mv = O(\lambda_L)$, as evidenced in (14.16). In 2D, this screening has no particular effect on the XY behaviour; in 3D, however, the effect is large. As for the confinement–deconfinement transition, it changes an otherwise 3D XY critical behaviour into the same Vogel–Fulcher–Tamman behaviour (14.48) [146]. 3D superconductors with emergent granularity and quasi-2D scaling behaviour have recently been experimentally detected [37]. At the time of writing, however, there are not yet any data on Vogel–Fulcher–Tamman critical behaviour of the superconducting transition.

Chapter 15

Gauging spin

The infrared-dominant term in the effective gauge action of the 3D SIT is the topological BF term (14.1). In this chapter we will ask the question of the microscopic origin of this term. In 2D, the topological Chern–Simons mass term is radiatively induced at one-loop level by coupling the gauge field to massive fermions, the induced coupling being determined by the sign of the fermion mass [147, 148]. As we now show, in 3D the BF term is correspondingly radiatively induced by gauging the spin of electrons [149].

We are thus looking for a conserved, antisymmetric two-tensor current of Dirac spinors that we can couple to an antisymmetric tensor gauge field $B_{\mu\nu}$. The obvious candidates are the tensor and pseudotensor currents $\bar\psi\sigma^{\mu\nu}\psi$ and $\bar\psi\gamma^5\sigma^{\mu\nu}\psi$, with $\sigma^{\mu\nu} \equiv (i/2)[\gamma^\mu, \gamma^\nu]$ and γ^μ the usual Dirac γ-matrices [150] ([,] denotes commutators and { , } anticommutators). Both, however, are not suitable since they are not conserved. Another candidate is the standard spin current of Dirac fermions, $S^{\mu,\alpha\beta} = (1/4)\bar\psi\{\gamma^\mu, \sigma^{\alpha\beta}\}\psi$. This, however, is a three-tensor. In order to obtain a two-tensor we shall consider the non-local field generated by the divergence of the spin current. Moreover, in order to maintain $\mathcal{P}$ and $\mathcal{T}$ conservation when coupling to the pseudotensor $B_{\mu\nu}$, we shall add a γ^5 matrix in the original relativistic spin current. We shall thus consider the coupling

$$\mathcal{L}_{\text{int}}^B = gB_{\mu\nu}J^{\mu\nu} \tag{15.1}$$

to the two-pseudotensor current

$$J^{\mu\nu} = \frac{-m}{2}\frac{\partial_\alpha}{\Box}\left(\bar\psi\gamma^5\{\gamma^\alpha, \sigma^{\mu\nu}\}\psi\right) = \frac{-im}{6}\frac{\partial_\alpha}{\Box}\left(\bar\psi\gamma^5\gamma^{[\alpha}\gamma^\mu\gamma^{\nu]}\psi\right), \tag{15.2}$$

where the symbol $[\ldots]$ denotes total antisymmetrization of the indices and g is a dimensionless coupling. This current is topologically conserved, as is evident from the second expression in (15.2). The coupling (15.1) is invariant under gauge transformations of the second kind,

$$B_{\mu\nu} \to B_{\mu\nu} + \partial_\mu \lambda_\nu - \partial_\nu \lambda_\mu, \tag{15.3}$$

and under the discrete parity and time-reversal transformations. Since the canonical dimension of fermion fields in 3D is 3/2, in this chapter we will take the antisymmetric tensor $B_{\mu\nu}$ as a field of canonical dimension 1, like the usual vector field A_μ.

In order to find out the nature of the coupling (15.1) we use the identity

$$\gamma^{[\alpha}\gamma^\mu \gamma^{\nu]} = 6i \, \epsilon^{\alpha\mu\nu\sigma} \gamma^5 \gamma_\sigma, \tag{15.4}$$

to rewrite the current $J^{\mu\nu}$ as

$$J^{\mu\nu} = m \, \epsilon^{\alpha\mu\nu\sigma} \frac{\partial_\alpha}{\Box} J_\sigma, \tag{15.5}$$

where $J^\mu = \bar{\psi}\gamma^\mu \psi$ is the usual Dirac current. Let us now focus on the "vector charges" $J^{0i} = -m \, \epsilon^{ijk}(\partial_j/\Box)\psi^\dagger \alpha^k \psi$. The expression $\psi^\dagger \alpha^k \psi$ is the velocity field of the Dirac fermion. Thus, in analogy to fluid dynamics, the pseudovector quantity $\Box J^{0i}$, given by the curl of the velocity field, represents the vorticity field of the fermions.

Using the Gordon decomposition [150]

$$\psi^\dagger \alpha^k \psi = \frac{i}{2m} \left[\bar{\psi}\partial^k \psi - \partial^k \bar{\psi}\psi \right] + \frac{1}{2m}\partial_\mu \, \bar{\psi}\sigma^{k\mu}\psi, \tag{15.6}$$

one can separate the velocity field into orbital and spin contributions. In the non-relativistic limit, in which the lower components of Dirac spinors can be neglected for energies much lower than their mass, this reduces to

$$(\psi^\dagger \alpha^k \psi)_{\rm NR} \to \frac{i}{2m} \left[\bar{\phi}\partial^k \phi - \partial^k \bar{\phi}\phi \right] + \frac{1}{2m}\epsilon^{kij}\partial_i \, \phi^\dagger \sigma^j \phi, \tag{15.7}$$

where the spinors ϕ on the right-hand side are the two-dimensional Pauli spinors corresponding to the upper components of the four-dimensional Dirac spinors ψ [150]. We obtain thus

$$J^{0i}_{\rm NR} = \Pi_{ij} \left(\phi^\dagger \frac{\sigma^j}{2} \phi \right), \tag{15.8}$$

where $\Pi_{ij} = (\delta_{ij} - \partial_j \partial_j/\Box)$ is the transverse projector. These are the transverse components of the electron magnetic moment density, which is why we call this a gauge theory of spin.

Let us consider the following field theory of electrons of mass m and charge e coupled to an electromagnetic field and to the antisymmetric tensor $B_{\mu\nu}$,

$$\mathcal{L} = \bar{\psi}\gamma^\mu\left(i\partial_\mu + eA_\mu\right)\psi - m\bar{\psi}\psi + gB_{\mu\nu}J^{\mu\nu} - \frac{1}{4}F_{\mu\nu}F^{\mu\nu} + \frac{1}{12}H_{\mu\nu\alpha}H^{\mu\nu\alpha},$$
(15.9)

where $H_{\mu\nu\alpha} \equiv \partial_{[\mu}B_{\nu\alpha]}$ is the three-tensor field strength associated to $B_{\mu\nu}$, as introduced in the preceding chapter.

By an integration by parts we can rewrite the antisymmetric tensor coupling as

$$\mathcal{L}^B_{\text{int}} = gB_{\mu\nu}J^{\mu\nu} = \frac{2mg}{\Box}F_\mu J^\mu,$$
(15.10)

where $F_\mu = (1/2)\epsilon_{\mu\nu\alpha\beta}\partial^\nu B^{\alpha\beta}$ is the dual field strength of $B_{\mu\nu}$. Formally, thus, the inverse d'Alembertian of the dual field strength plays the role of a gauge field in the Lorenz gauge and we can rewrite the total interaction term as

$$\mathcal{L}_{\text{int}} = e(A_{\text{eff}})_\mu J^\mu,$$
$$(A_{\text{eff}})_\mu = A_\mu + \frac{2mg}{e\Box}F_\mu.$$
(15.11)

In order to integrate out the fermions and obtain the one-loop induced action we need now only the standard vacuum polarisation tensor. In momentum space this is given by

$$\Gamma^{(1-\text{loop})} = -\frac{ie^2}{2}\text{Tr}\left(\frac{1}{i\slashed{\partial} - m}\slashed{A}_{\text{eff}}\frac{1}{i\slashed{\partial} - m}\slashed{A}_{\text{eff}}\right)$$
$$= \frac{1}{2}\int\frac{d^4k}{(2\pi)^4}(A_\mu)_{\text{eff}}(-k)(A_\nu)_{\text{eff}}(k)\Pi^{\mu\nu}(k)$$
(15.12)

where $\Pi^{\mu\nu}$ is the usual quantum electrodynamics (QED) vacuum polarization tensor [150]

$$\Pi^{\mu\nu}(k) = (g^{\mu\nu}k^2 - k^\mu k^\nu)\Pi(k^2),$$
$$\Pi(k^2) = \frac{e^2}{12\pi^2}\ln\frac{\Lambda^2}{m^2},$$
(15.13)

and Λ is the momentum-space UV cutoff.

We obtain thus the one-loop induced action

$$\Gamma^{1-\rm loop} = \frac{e^2}{4}\frac{1}{12\pi^2}\ln\frac{\Lambda^2}{m^2}\int d^4x \ (F_{\rm eff})_{\mu\nu}(F_{\rm eff})^{\mu\nu}, \tag{15.14}$$

with

$$(F_{\rm eff})_{\mu\nu} = F_{\mu\nu} + \frac{2mg}{e}\left(\frac{\partial_\mu}{\Box}F_\nu - \frac{\partial_\nu}{\Box}F_\mu\right). \tag{15.15}$$

Expanding this expressions leads to three terms in the original variables

$$\Gamma^{1-\rm loop} = \Gamma^{1-\rm loop}_{AA} + \Gamma^{1-\rm loop}_{AB} + \Gamma^{1-\rm loop}_{BB}, \tag{15.16}$$

where

$$\Gamma^{1-\rm loop}_{AA} = \frac{e^2}{4}\frac{1}{12\pi^2}\ln\frac{\Lambda^2}{m^2}\int d^4x \ F_{\mu\nu}F^{\mu\nu}, \tag{15.17}$$

$$\Gamma^{1-\rm loop}_{AB} = \frac{8mg}{4}\frac{1}{12\pi^2}\ln\frac{\Lambda^2}{m^2}\int d^4x \ A_\mu F^\mu, \tag{15.18}$$

$$\Gamma^{1-\rm loop}_{BB} = \frac{8m^2g^2}{4}\frac{1}{12\pi^2}\ln\frac{\Lambda^2}{m^2}\int d^4x \ F_\mu\frac{1}{\Box}F^\mu. \tag{15.19}$$

We can then combine this induced action with the original one to obtain the one-loop effective Lagrangian. With the usual rescaling $A_\mu \to A_\mu/e$, to bring the action to its standard form, we obtain

$$\mathcal{L} = -\frac{1}{4e_{\rm ph}^2}F_{\mu\nu}F^{\mu\nu} + \frac{g_{\rm ph}m}{2\pi}B_{\mu\nu}\epsilon^{\mu\nu\rho\sigma}\partial_\rho A_\sigma + \frac{1}{12}H_{\mu\nu\alpha}H^{\mu\nu\alpha}$$

$$+ \frac{6g_{\rm ph}^2 m^2}{\ln\frac{\Lambda^2}{m^2}}F_\mu\frac{1}{\Box}F^\mu, \tag{15.20}$$

where

$$e_{\rm ph}^2 = e^2\left(1 + \frac{e^2}{12\pi^2}\ln\frac{\Lambda^2}{m^2}\right) \tag{15.21}$$

is the QED renormalized charge and

$$g_{\rm ph} = \frac{g}{6\pi}\ln\frac{\Lambda^2}{m^2}. \tag{15.22}$$

As anticipated, the topological BF mass term is induced at one-loop level. In general, however, this is a non-renormalizable model, valid only at momentum scales much smaller than the cutoff. Things are different, however, if we consider the fermion mass as an infrared (IR) regulator to be removed at the end of the computation. In this case, by choosing $\Lambda = m \, \exp(\mu/2m)$ we can simultanously remove the IR and UV regulators so that

$$\lim_{\substack{m \to 0 \\ \Lambda \to \infty}} m \, \ln\frac{\Lambda^2}{m^2} = \mu. \tag{15.23}$$

In this limit the induced BF mass term is finite, while the coefficient of the non-local BB one-loop-induced term vanishes, as does the Maxwell term.

Chapter 16

Dyons and the θ-term

We have established that, under certain conditions, the electromagnetic response of insulating materials can become the compact version of QED [18, 19], which admits magnetic monopoles. When these Bose condense, superinsulation is induced and electric charge is confined. In this chapter we shall, instead, consider the physics of non-condensed magnetic monopoles, in particular their interaction with non-confined electric charge.

Let us thus consider a bound state of a monopole of magnetic charge g and a particle of electric charge e. Such objects, carrying both electric and magnetic charge, are called *dyons* (for a review see [17]). To make things simple, we consider an infinitely heavy magnetic monopole fixed at the origin and a light particle of charge e and mass m moving in the radial magnetic field $\mathbf{B} = (g/4\pi r^2)\hat{\mathbf{r}}$, where $\hat{\mathbf{r}}$ is the unit vector in direction $\mathbf{r}$. This magnetic field is spherically symmetric and, thus, one would expect conservation of angular momentum. However, in this case, the Lorentz force entering the equation of motion

$$m\frac{d^2}{dt^2}\mathbf{r} = e\frac{d}{dt}\mathbf{r} \wedge \mathbf{B}, \tag{16.1}$$

is not a central force. As a consequence, it is not the usual orbital angular momentum which is conserved,

$$\frac{d}{dt}\left(\mathbf{r} \wedge m\frac{d}{dt}\mathbf{r}\right) = \mathbf{r} \wedge m\frac{d^2}{dt^2}\mathbf{r} = \frac{eg}{4\pi r^3}\mathbf{r} \wedge \left(\frac{d}{dt}\mathbf{r} \wedge \mathbf{r}\right) = \frac{d}{dt}\left(\frac{eg}{4\pi}\hat{\mathbf{r}}\right). \tag{16.2}$$

The conserved total angular momentum is, rather,

$$\mathbf{J} = \mathbf{r} \wedge m\frac{d}{dt}\mathbf{r} - \frac{eg}{4\pi}\hat{\mathbf{r}}. \tag{16.3}$$

147

What is the origin of this additional contribution? To establish this, let us consider the angular momentum of the electromagnetic field,

$$\mathbf{J}_{\text{e.m.}} = \int d^3\mathbf{x} \; \mathbf{x} \wedge (\mathbf{E} \wedge \mathbf{B}), \qquad (16.4)$$

for the case in which $\mathbf{B}$ is the radial field of a magnetic charge g at the origin and $\mathbf{E}$ the field of an electric charge at $\mathbf{r}$. This gives

$$\begin{aligned}
J^i_{\text{e.m.}} &= \int d^3\mathbf{x} \; \frac{g}{4\pi|x|} E^i \left(\delta^{ij} - \hat{x}^i \hat{x}^j\right) \\
&= \int d^3\mathbf{x} \; E^j \frac{\partial}{\partial x^j} \frac{g\hat{x}^i}{4\pi} \\
&= -\int d^3\mathbf{x} \frac{g}{4\pi} \nabla \cdot \mathbf{E} \; \hat{x}^i \\
&= -\frac{eg}{4\pi} \hat{\mathbf{r}}^i,
\end{aligned} \qquad (16.5)$$

where we have used $\nabla \cdot \mathbf{E} = e\delta^3(\mathbf{x} - \mathbf{r})$. The conserved angular momentum is thus the sum of the orbital angular momentum of the particle and the angular momentum of the electromagnetic field.

This classical result implies two very important consequences when we "switch on" quantum mechanics. First, the non-Abelian character of the rotation group in 3D implies that angular momentum is quantized in integer multiples of $(1/2)$. Hence we obtain the requirement

$$\frac{eg}{4\pi} = \frac{n}{2}, \qquad n \in \mathbb{Z}, \qquad (16.6)$$

which is nothing else than the same Dirac quantization condition we derived in Chapter 2 by the requirement that the Dirac string must be unobservable. This very strong compatibility condition with the quantum mechanical angular momentum quantization shows that true magnetic monopoles, even emergent ones in condensed matter systems, must satisfy the Dirac quantization condition, contrary to the endpoints of magnetic dipoles, even if these are very long.

Unit dyons with $eg = 2\pi$ have spin $1/2$. The usual spin-statistics connection would thus imply that they are fermions and that monopoles induce statistical transmutation [151]. To show that this is indeed so, we consider the Hamiltonian for two identical dyons of mass m, electric charge e, and magnetic charge g:

$$H = \frac{1}{2m} \left(\mathbf{p}_1 - e\mathbf{A}(\mathbf{x}_1 - \mathbf{x}_2)\right)^2 + \frac{1}{2m} \left(\mathbf{p}_2 - e\mathbf{A}(\mathbf{x}_2 - \mathbf{x}_1)\right)^2 + V_C \left(|\mathbf{x}_1 - \mathbf{x}_2|\right), \tag{16.7}$$

where $\mathbf{A}$ is the monopole gauge potential and $V_C(r) = e^2/4\pi r$ is the Coulomb potential. The first two terms in this Hamiltonian represent the interactions of the electric charge e at $\mathbf{x}_1$ with the magnetic charge g at $\mathbf{x}_2$ and *vice versa*. We introduce, as usual, centre-of-mass and relative coordinates $\mathbf{R} = (\mathbf{x}_1 + \mathbf{x}_2)/2$ and $\mathbf{r} = (\mathbf{x}_1 - \mathbf{x}_2)$ and we set $\mathbf{R} = 0$. The time-independent Schrödinger equation becomes

$$\left[\frac{-1}{2m} \left(\nabla - ie\mathbf{A}\left(\mathbf{r}\right)\right)^2 + \frac{-1}{2m} \left(\nabla + ie\mathbf{A}\left(-\mathbf{r}\right)\right)^2 + V_C(|\mathbf{r}|)\right] \psi = E\psi. \tag{16.8}$$

Now comes a crucial point: if one dyon is in the upper hemisphere, in these coordinates, the other one must be in the lower hemisphere. If we restrict ourselves to the monopole vector potential that we introduced in Chapter 2,

$$\mathbf{A}_u = f_u(r,\theta)\,\hat{\varphi},$$
$$f_u(r,\theta) = \frac{g}{4\pi r}\frac{1 - \cos(\theta)}{\sin(\theta)}, \tag{16.9}$$

where r, θ, and φ denote the usual spherical coordinates and $\hat{\varphi}$ is the unit vector in φ direction, the second dyon in the lower hemisphere will feel a singular gauge configuration at the south pole $\theta = \pi$. To solve problems in the field of a Dirac monopole one cannot use a single set of coordinates for the whole sphere but one must use the Wu–Yang formalism [152] and cover the sphere with an atlas of maps, supplemented by gauge transformation conditions between wave functions in the overlap regions of different maps. The simplest atlas consists of two maps, the northern hemisphere $0 \le \theta \le \pi/2 + \epsilon$, with the gauge potential (16.9), and the lower hemisphere $\pi/2 - \epsilon \le \theta \le \pi$, with the gauge transformed gauge potential

$$\mathbf{A}_l = f_l(r,\theta)\,\hat{\varphi},$$
$$f_l(r,\theta) = -\frac{g}{4\pi r}\frac{1 + \cos(\theta)}{\sin(\theta)}, \tag{16.10}$$

corresponding to the same magnetic monopole at the origin but with the Dirac string now along the positive z-axis,

$$\mathbf{A}_l = \mathbf{A}_u - \nabla\left(\frac{g}{2\pi}\varphi\right). \tag{16.11}$$

In both hemispheres the gauge potential is now regular and the corresponding Schrödinger equation can be solved. The price to pay are gauge transformation conditions between the wave functions on the overlap $[\pi/2 - \epsilon < \theta < \pi/2 + \epsilon]$ between the two hemispheres,

$$\psi_l(r, \theta, \varphi) = \mathrm{e}^{-i\frac{eg}{2\pi}\varphi}\psi_u(r, \theta, \varphi). \tag{16.12}$$

Using the fact that the monopole gauge potentials (16.9) and (16.10) are divergenceless, we obtain

$$\left[\frac{-1}{m}\nabla^2 + \frac{ie}{2m}\left(\mathbf{A}\left(\mathbf{r}\right) \cdot \nabla - \mathbf{A}\left(-\mathbf{r}\right) \cdot \nabla\right)\right.$$
$$\left.+ \frac{e^2}{2m}\left(\mathbf{A}^2\left(\mathbf{r}\right) + \mathbf{A}^2\left(-\mathbf{r}\right)\right) + V_C(|\mathbf{r}|)\right]\psi = E\psi. \tag{16.13}$$

Of course, this is a formal expression. To be precise we have to formulate two Schrödinger equations, one for the upper and one for the lower hemisphere, by using the corresponding gauge potentials (16.9) and (16.10). In doing so we will neglect the third term and the Coulomb potential, which are irrelevant for what follows. Using the expressions (16.9) and (16.10) gives

$$\left[\frac{-1}{m}\nabla^2 + i\frac{(eg/2\pi)}{mr^2}\frac{1 - \cos(\theta)}{\sin^2(\theta)}\frac{\partial}{\partial\varphi}\right]^2 \psi_u = E\psi_u,$$
$$\left[\frac{-1}{m}\nabla^2 - i\frac{(eg/2\pi)}{mr^2}\frac{1 + \cos(\theta)}{\sin^2(\theta)}\frac{\partial}{\partial\varphi}\right]^2 \psi_l = E\psi_l. \tag{16.14}$$

Let us now investigate the symmetry of the wave function under exchange of the two dyons, which corresponds to the exchange $\mathbf{r} \leftrightarrow -\mathbf{r}$. In centre-of-mass spherical coordinates, this corresponds to $(r, \theta, \varphi) \leftrightarrow (r, \pi - \theta, \varphi + \pi)$. But this involves unavoidably an exchange of hemispheres. Therefore, the exchange must involve a gauge transformation too. In the overlap region, where both wave functions are defined,

$$\psi_u(r, \theta, \varphi) \to \mathrm{e}^{+i\frac{eg}{2\pi}\varphi}\psi_l(r, \pi - \theta, \varphi + \pi) = \mathrm{e}^{-i\frac{eg}{2\pi}\pi}\psi_u(r, \pi - \theta, \varphi + \pi),$$
$$\psi_l(r, \theta, \varphi) \to \mathrm{e}^{-i\frac{eg}{2\pi}\varphi}\psi_u(r, \pi - \theta, \varphi + \pi) = \mathrm{e}^{+i\frac{eg}{2\pi}\pi}\psi_l(r, \pi - \theta, \varphi + \pi).$$

$$\tag{16.15}$$

Let us assume that charges are bosons. The wave functions of unit dyons with $eg/2\pi = 1$ are then odd under exchange, which confirms that such dyons are indeed *fermions* and that the usual spin-statistics relation is maintained. In general, the statistics of a composite of n bosonic electric charges and m magnetic monopoles is $(-1)^{nm}$, where $+1$ corresponds to bosons and -1 to fermions.

Equation (16.15) exposes another important consequence of the additional spin (16.3) of dyons. At the quantum level, only the z-component of the angular momentum is an observable. Because the monopole vector potential breaks rotational invariance, the z-component from the dyon spin explicitly appears in the Schrödinger equation, where it can modify the usual centrifugal barrier. This has very important consequences, to be discussed in a later chapter.

Is it possible to have dyons with no statistical transmutation? The answer is affirmative [153], but before describing how these fundamental dyons (as opposed to the composite ones discussed above) arise we need to introduce some more topology. In 3D, the Maxwell action (2.22) is not the only Lorentz-invariant and gauge-invariant action one can write down for electromagnetic fields. An additional possible term with a dimensionless coupling is the so-called θ-term,

$$S_\theta = \frac{\theta}{16\pi^2} \int d^4x \, \tilde{F}_{\mu\nu} F^{\mu\nu} = \frac{\theta}{32\pi^2} \int d^4x \, F_{\mu\nu} \epsilon^{\mu\nu\alpha\beta} F_{\alpha\beta} = \frac{\theta}{4\pi^2} \int d^4x \, \mathbf{E} \cdot \mathbf{B},$$

$$(16.16)$$

where $\tilde{F}^{\mu\nu} = (1/2)\epsilon^{\mu\nu\alpha\beta} F_{\alpha\beta}$ is the dual electromagnetic tensor. When θ is a constant, this term represents the axial anomaly, responsible for the $\pi^0 \to 2\gamma$ decay of the neutral pion [40]. When $\theta(x)$ is a field, it represents an additional particle coupled to electromagnetic fields, the axion. The correspondingly modified electromagnetic theory is called axion electrodynamics [154].

As the 2D Chern–Simons term, the 3D θ-term is topological, since it can be defined independently of a space-time metric. Contrary to the 2D Chern–Simons term, it reduces to a pure boundary contribution on bounded (Euclidean) manifolds M,

$$S_\theta = \frac{\theta}{8\pi^2} \int_M d^4x \, \partial_\mu \left(A_\nu \epsilon^{\mu\nu\alpha\beta} \partial_\alpha A_\beta \right) = \frac{(\theta/2\pi)}{4\pi} \int_{\partial M} d^3x \, A_\mu \epsilon^{\mu\alpha\nu} \partial_\alpha A_\nu,$$

$$(16.17)$$

where ∂M represents the boundary of M. This shows that a bulk θ-term is entirely equivalent to a boundary Chern–Simons term on bounded manifolds. The θ-term involves the scalar product of a vector field, $\mathbf{E}$, and a pseudovector field, $\mathbf{B}$. Therefore it is odd under parity and time-reversal. Actually, as we now show, this is not entirely true for compact gauge fields since, for particular values of θ, the quantity $\exp(iS_\theta)$ entering the partition function is even under both $\mathcal{P}$ and $\mathcal{T}$.

When the gauge group is non-compact, the θ-term can be any real number. Things change dramatically, however, for compact gauge groups. As stressed by Polyakov [19], there are two ways to obtain a compact U(1) theory: either by spontaneously breaking a compact non-Abelian gauge group G down to U(1) by some Higgs fields or by formulating the model on a lattice. In the following we shall follow the first avenue. To be specific, we will choose $G = \mathrm{SU}(2)$ and consider the Euclidean, non-Abelian θ-term (note the additional factor "i" on Euclidean manifolds)

$$S_\theta = \frac{i\theta}{16\pi^2} \int d^4x \ \mathrm{Tr} \ \tilde{F}_{\mu\nu} F^{\mu\nu}, \qquad (16.18)$$

where

$$F_{\mu\nu} = F^a_{\mu\nu} T^a, \qquad (16.19)$$

and the anti-Hermitean generators T^a of the non-Abelian group, with commutation relations

$$[T^a, T^b] = f^{abc} T^c, \qquad (16.20)$$

and structure constants f^{abc} are normalized as

$$\mathrm{Tr} \ T^a T^b = -\frac{1}{2}\delta^{ab}. \qquad (16.21)$$

The non-Abelian (Euclidean) version of (16.17) is

$$S_\theta = \frac{i\theta}{16\pi^2} \int_{S^3} d^3x \ \mathrm{Tr} \ \epsilon_{ijk} \left(A_i F_{jk} - \frac{2}{3} A_i A_j A_k \right), \qquad (16.22)$$

where the integral is taken over a three-sphere S^3 at infinity in Euclidean space-time [20, 40]. The boundary conditions require that physical electromagnetic fields F_{ij} vanish at infinity. This, however, does not imply that also the gauge potentials must vanish there. These are just restricted to be pure gauge configuration $A_i = g^{-1}\partial_i g$ on the sphere S^3 at infinity,

$$S_\theta = -\frac{i\theta}{24\pi^2} \int_{S^3} d^3x \ \mathrm{Tr} \ \epsilon_{ijk} \left(g^{-1}\partial_i g \ g^{-1}\partial_j g \ g^{-1}\partial_k g \right). \qquad (16.23)$$

Since we have spontaneously broken SU(2) to U(1), the gauge group elements g are defined only up to the residual U(1) gauge transformations, i.e. $g \in \mathrm{SU}(2)/\mathrm{U}(1)$. The group SU(2) is homeomorphic to the three-sphere

S^3, while U(1) to the circle S^1. The coset SU(2)/U(1), the so-called Hopf bundle (for a review see [155]), is thus homeomorphic to the sphere S^2. The θ-term defines thus a mapping from the sphere S^3 to the sphere S^2. These mappings are classified by the third homotopy group $\pi^3\left(S^2\right) = \mathbb{Z}$. Specifically

$$S_\theta = i\theta n, \tag{16.24}$$

where n is an integer. Since only the exponential $\exp\left(-S_\theta\right)$ in the partition function matters, we obtain the result that $\theta \in [0, 2\pi]$ is an *angle*. Moreover, since $\exp\left(-S_\pi\right) = \exp\left(-S_{-\pi}\right)$, neither the parity nor the time-reversal symmetries are broken for $\theta = \pi$ mod 2π. We have derived these results by specializing to the spontaneous breaking of SU(2) to U(1). However, they turn out to be valid in general.

The θ-term has a very important consequence for what follows. Let us integrate by parts in (16.16). This gives

$$S_\theta = \frac{\theta}{2\pi} \int d^4x \, A_\mu j_g^\mu,$$

$$j_g^\mu = -\frac{1}{4\pi}\epsilon^{\mu\nu\alpha\beta}\partial_\nu F_{\alpha\beta} = -\frac{1}{2\pi}\partial_\nu \tilde{F}^{\mu\nu}. \tag{16.25}$$

For non-compact gauge fields, j_g^μ vanishes identically. For compact U(1) gauge fields, instead, singularities are admitted and j_g^μ is the corresponding magnetic monopole current introduced in Chapter 14 (note that $2\pi j_g^0 = \nabla\cdot\mathbf{B}$). As a consequence of the θ-term, magnetic monopoles become dyons carrying also electric charge $\theta/2\pi$. This topologically acquired electric charge is called the Witten effect [156]. Such dyons are excited states of magnetic monopoles and they are thus *fundamental dyons*, different from the *composite dyons* we have introduced above. Correspondingly, they remain bosons, no statistical transmutation is induced by the θ-term [153].

The θ-term is not only a theoretical fluke of gauge theories. There are concrete materials which realize this topological term in their electromagnetic response. These are called *strong* topological insulators [99] and are characterized by the value $\theta = \pi$. Since time-reversal symmetry is present only for the values $\theta = 0$ and $\theta = \pi$, no perturbation which preserves time-reversal symmetry can turn a strong topological insulator into a normal one. The θ-term leads thus to a topological $\mathbb{Z}_2$ index characterizing insulators.

The consequences of the presence of a θ-term in the electromagnetic response are felt mostly at the surfaces of topological insulators. To see this, let us consider the equations of motion of axion electrodynamics, with a spatially varying field $\theta(\mathbf{x})$,

$$\nabla \cdot \mathbf{E} = -\frac{e^2}{4\pi^2} \nabla\theta \cdot \mathbf{B},$$

$$-\frac{\partial \mathbf{E}}{\partial t} + \nabla \wedge \mathbf{B} = \frac{e^2}{4\pi^2} \nabla\theta \wedge \mathbf{E},$$

$$\nabla \cdot \mathbf{B} = 0, \tag{16.26}$$

$$\frac{\partial \mathbf{B}}{\partial t} + \nabla \cdot \mathbf{E} = 0.$$

We will now fill the lower half-space $z < -\epsilon$ with a topological insulator with $\theta = \pi$ and the upper half-space $z > +\epsilon$ with the vacuum, having $\theta = 0$. If we apply a uniform magnetic field B_z in the topological insulator, the first equation in (16.26) tells us that the change of θ at the interface between $-\epsilon$ and $+\epsilon$ corresponds to a charge density on the surface. This will generate an electric field E_z in the vacuum above. Alternatively, if we apply a static horizontal electric field, say E_x, in the topological insulator, the second equation in (16.26) tells us that the change of θ at the interface corresponds to a Hall current in y-direction, which will generate a magnetic field in the (x, z) plane in the vacuum above. This phenomenon of magnetic fields generating electric ones and *vice versa* is called the *magnetoelectric effect* [157], the optical Kerr effect being one of its manifestations [158].

Strong bosonic topological insulators [136] can be obtained by adding the θ-term for the emergent gauge field a_μ to their effective field theory (14.44),

$$\mathcal{L}_{\text{strong}} = \frac{1}{4\pi} a_\mu \epsilon^{\mu\nu\alpha\beta} \partial_\nu b_{\alpha\beta} + \frac{\theta}{8\pi^2} \epsilon^{\mu\nu\alpha\beta} \partial_\mu a_\nu \partial_\alpha a_\beta. \tag{16.27}$$

In this case, however, the periodicity is doubled to $\theta \to \theta + 4\pi$ and the value characterizing strong bosonic topological insulators is $\theta = 2\pi$. We will see in a later chapter why this is the case.

Chapter 17

The electric Meissner effect, antiscreening and confining strings

The electromagnetic response of Mott insulators is the usual Maxwell action, augmented by the θ-term for strong topological insulators, that of superconductors is the Proca action, including a photon mass term $\lambda_L^{-2} A_\mu A^\mu$ which, in the physical Weyl gauge $A_0 = 0$, is equivalent to the London equations $\mathbf{j}_{\mathrm{ind}} = \lambda_L^{-2} \mathbf{A}$. What is the corresponding electromagnetic response of superinsulators? Until now we have derived that, in the superinsulating phase, positive and negative charges are linearly bound by an electric flux string, which causes the infinite resistance. In this chapter we will focus on the full electromagnetic response and we will derive the field theory for the confining strings.

As we have explained in Sec. 14.3, the electromagnetic response of superinsulators consists of two parts, the usual insulator part, leading to a sub-dominant screened Coulomb potential, and the "super" part, arising from the condensation of magnetic monopoles and giving rise to the dominant linear potential. In the following we shall neglect the first and focus only on the monopole-induced part. The question we address is whether the monopole condensate can be represented as an effective field theory coupled to the electric strings causing the linear potential.

To answer this question we follow [159] and start from the Euclidean space-time representation (14.35) of the fundamental monopole condensate, where we omit, for the moment, the Wilson loop, we reinstate the explicit lattice spacing $\ell = 1/\Lambda_0$, with Λ_0 the UV cutoff, and we introduce the

155

simple notation e^2 for the combination $f^2/8\pi^2$,

$$Z_\ell = \int \mathcal{D}\varphi_\mu \, \exp\left(-S_\ell\right),$$

$$S_\ell = \sum_x \left(4e^2\ell^4 f_{\mu\nu}f_{\mu\nu} + z\left(1 - \cos(\ell\varphi_\mu)\right)\right). \tag{17.1}$$

We recall that z represents the fugacity of the monopole plasma. We now introduce an antisymmetric tensor field $B_{\mu\nu}$ by representing (17.1) as,

$$Z_\ell = \int \mathcal{D}B_{\mu\nu}\mathcal{D}\varphi_\mu \, \exp\left(-S(B_{\mu\nu}, \varphi_\mu)\right), \tag{17.2}$$

with the first-order Euclidean action

$$S_\ell = S(B_{\mu\nu}, \varphi_\mu)$$

$$= \sum_x \ell^4 \left[\frac{1}{4e^2}B_{\mu\nu}\dot{B}_{\mu\nu} + i\varepsilon_{\mu\nu\alpha\beta}B_{\mu\nu}f_{\alpha\beta} + z\frac{1}{\ell^4}\left(1 - \cos\left(\ell\varphi_\mu\right)\right)\right]. \tag{17.3}$$

Gaussian integration over $B_{\mu\nu}$ reproduces (17.1). We are interested, however, to find the dual action in terms of $B_{\mu\nu}$ alone, by integrating over φ_μ. Notice that the non-linear dependence on this field makes the integration seem essentially untractable. However, one can proceed as follows. We first rewrite (17.2) and (17.3) as

$$Z_\ell = \int \mathcal{D}B_{\mu\nu} \sum_{\{n_\mu\}} \exp\left(2\pi i \sum_x \ell^3 B_{\mu\nu}t_{\mu\nu}\right)$$

$$\times \int_{-\pi/\ell}^{+\pi/\ell} \mathcal{D}\varphi_\mu \, \exp\left(-S_\ell\left(B_{\mu\nu}, \varphi_\mu\right)\right), \tag{17.4}$$

with $t_{\mu\nu} = \epsilon_{\mu\nu\alpha\beta}\left(d_\alpha n_\beta - d_\beta n_\alpha\right)$. In doing so, we have restricted the φ_μ integrations to the fundamental domain, at the price of introducing an additional set of integer link variables n_μ which take into account all other periods of the cos function. At this point we can perform the integrations over φ_μ by exploiting the formula

$$e^{a\cos x} = \sum_{k \in \mathbb{Z}} I_k(a)\, e^{ikx}, \tag{17.5}$$

with $I_k(a)$ a modified Bessel function of the second kind. Using this we obtain the following contribution to the partition function from the φ_μ

integrations (up to an irrelevant factor):

$$
Z_\varphi = \prod_{x,\mu} \int_{-\pi/\ell}^{+\pi/\ell} d\varphi_\mu(x) \; \exp\left(z\cos(\ell\varphi_\mu) + i\frac{2}{3}l^4\varphi_\mu H_\mu \right)
$$

$$
= \prod_{x,\mu} \sum_{n_\mu(x)} I_{n_\mu}(z) \int_{-\pi}^{+\pi} d\varphi_\mu \; \exp i\left(\frac{2}{3}\ell^3\varphi_\mu H_\mu - n_\mu\varphi_\mu \right)
$$

$$
= \prod_{x,\mu} \sum_{n_\mu(x)} I_{n_\mu}(z)\, \delta_{n_\mu,\frac{2}{3}\ell^3 H_\mu}
$$

$$
= \prod_{x,\mu} I_{\frac{2}{3}\ell^3 H_\mu}(z), \tag{17.6}
$$

where we have absorbed ℓ in the definition of φ_μ and $H_\mu = (1/6)\epsilon^{\mu\nu\alpha\beta}H_{\nu\alpha\beta} = (1/2)\epsilon^{\mu\nu\alpha\beta}\partial_\nu B_{\alpha\beta}$ is the dual field strength associated to the antisymmetric tensor, as already introduced in Chapter 14. Note that the Kronecker delta conditions imply the quantization condition $(2/3)\ell^3 H_\mu(x) \in \mathbb{Z}$ for all x and μ. For strong coupling e (large monopole fugacity), the angles φ_μ are peaked around the maxima of the cosine function and, correspondingly, large values of the dual field $B_{\mu\nu}$ are favoured in the partition function. In this case, only small values $|\ell t_{\mu\nu}|$ contribute and we shall thus restrict to $|\ell t_{\mu\nu}| = \pm 1$. In this "dilute gas approximation" the sum over configurations $\{n_\mu\}$ reduces to a sum over *closed surfaces* on the lattice. Restoring the continuum notation by taking into account the appropriate factors of the UV cutoff Λ_0 we obtain

$$
Z = \int_{\substack{\text{closed} \\ \text{surfaces}}} \int \mathcal{D}B_{\mu\nu} \exp\left(-S\left(B_{\mu\nu}\right) + i\int_{\text{surface}} B_{\mu\nu}d\sigma_{\mu\nu} \right),
$$
$$
S\left(B_{\mu\nu}\right) = \int d^4x \left(-\Lambda_0^4 \log I_{\left(\frac{2H_\mu}{3\Lambda_0^3}\right)}(z) + \frac{1}{4e^2}B_{\mu\nu}B_{\mu\nu} \right), \tag{17.7}
$$

where $d\sigma_{\mu\nu}$ is the area element of the world-surface of a string. That is, we obtain the partition function of a *string theory* with action induced by the Kalb–Ramond tensor field $B_{\mu\nu}$.

In the semiclassical approximation, the modified Bessel function $I_p(a)$ satisfies

$$
\ln I_p(a) \sim (p^2 + a^2)^{1/2} - p\sinh^{-1}(p/a), \tag{17.8}
$$

and, for small p/a,

$$\frac{I_p(a)}{I_0(a)} \sim e^{-p^2/2a}. \tag{17.9}$$

Therefore, for $H_\mu \ll \Lambda_0^3$, we obtain the local partition function

$$Z = \int_{\substack{\text{closed} \\ \text{surfaces}}} \int \mathcal{D}B_{\mu\nu} \exp\left(-S\left(B_{\mu\nu}\right) + i \int_{\text{surface}} B_{\mu\nu} d\sigma_{\mu\nu}\right),$$

$$S\left(B_{\mu\nu}\right) = \int d^4x \left(\frac{1}{12\Lambda^2} H_{\mu\nu\alpha} H_{\mu\nu\alpha} + \frac{1}{4e^2} B_{\mu\nu} B_{\mu\nu}\right), \tag{17.10}$$

(after the overall factor $I_0(a)$ is absorbed in the integral), where

$$\Lambda = \frac{\Lambda_0}{4}\sqrt{z}. \tag{17.11}$$

This is the action of the *confining string*, first posited in [160] based on symmetry considerations, then derived in Minkowski space-time in [161] and rederived in Euclidean space from the monopole plasma in [159]. The key feature of the dynamical generation of strings in this model is the periodic structure of the potential induced by the condensation of the magnetic monopoles. Strings arise from the summation over the periods of this potential in the duality transformation. The summation is over closed surfaces, representing non-perturbative pure gauge excitations, the analogue of glueballs in quantum chromodynamics [4]. When a Wilson loop is included in the computation, the sum extends over all surfaces bounded by the contour of the Wilson loop. These are the fluctuating world-surfaces of the open strings between the positive and negative charges at the endpoints. Note that, in the limit $\Lambda \to 0$ of vanishing monopole fugacity, the action (17.10) reduces to the action of QED, since only configurations for which $H_{\mu\nu\alpha} = 0$ contribute to the partition function in this case. These are pure gauge configurations for which $B_{\mu\nu} = F_{\mu\nu} = \partial_\mu A_\nu - \partial_\nu A_\mu$. Correspondingly, the Wilson loop measures the Coulomb interaction of the endpoints of open strings,

$$\lim_{\Lambda \to 0} \left(-\ln\langle W(C)\rangle\right) = \frac{e^2}{2} \int d^4x \; j_\mu \frac{1}{-\nabla^2} j_\mu,$$

$$j_\mu(x) = 2\partial_\nu T_{\mu\nu}(x) = \int_C d\tau \; \frac{dx_\mu}{d\tau} \; \delta^4\left(x - x(\tau)\right), \tag{17.12}$$

where C denotes the closed world-line bounding the original open world-sheet.

The action in (17.10) is the low-energy electromagnetic response of superinsulators. There are no gauge potentials A_μ anymore, the condensation of magnetic monopoles causes the original electric and magnetic fields $\mathbf{E}$ and $\mathbf{B}$, encoded in the tensor $F_{\mu\nu}$, to be promoted themselves to the role of fundamental degrees of freedom, whose dynamics is encoded in the three-tensor $H_{\mu\nu\alpha}$ made of their derivatives [160]. Correspondingly, there are no point charges coupling to A_μ anymore, these are confined and only neutral strings coupling directly to $\mathbf{E}$ and $\mathbf{B}$ survive. As in superconductors, one additional degree of freedom has emerged, since the dynamical fields B_{ij} (B_{0i} are Lagrange multipliers) contain three degrees of freedom. However, this is not the Goldstone mode representing phase fluctuations of a superconducting parameter. It represents, rather, long-wavelength fluctuations of the magnetic monopole condensate. As we now show, this additional degree of freedom also gives mass to the photon. Contrary to the usual Anderson–Higgs mechanism, however, it is not the photon that "eats up" the Goldstone boson but the other way round, it is the new mode that absorbs the two transverse photon degrees of freedom to become a massive vector in a dual, generalized Stückelberg mechanism [160].

The Minkowski space-time Lagrangian corresponding to (17.10) is

$$\mathcal{L} = \frac{1}{12\Lambda^2} H_{\mu\nu\alpha} H^{\mu\nu\alpha} - \frac{1}{4e^2} B_{\mu\nu} B^{\mu\nu}. \tag{17.13}$$

The equations of motion of this model are

$$\partial_\mu H^{\mu\alpha\beta} + \frac{\Lambda^2}{e^2} B^{\alpha\beta} = 0. \tag{17.14}$$

Contracting with ∂_α we obtain the condition

$$\partial_\mu B^{\mu\nu} = 0. \tag{17.15}$$

Finally, contracting (17.14) with $\epsilon_{\nu\gamma\alpha\beta}\partial^\gamma$ and using the above condition we get

$$\left(\Box^2 + m^2\right) H^\mu = 0,$$
$$m = \frac{\Lambda}{e} = \frac{\Lambda_0\sqrt{z}}{4e}. \tag{17.16}$$

Because of the condition (17.15) we have

$$\partial_\alpha \left(\partial_\mu B^{\mu\beta}\right) - \partial_\beta \left(\partial_\mu B^{\mu\alpha}\right) = 0. \tag{17.17}$$

Using this in (17.14) we obtain the same massive equation at the level of electromagnetic fields,

$$\left(\Box^2 + m^2\right) B^{\mu\nu} = 0. \tag{17.18}$$

This shows that the monopole condensate indeed generates a mass for electromagnetic fields.

Let us now consider a slab of superinsulator occupying the $z > 0$ upper half-plane and let us apply a static, uniform electric field E_{ext} in the x-direction. For static fields, the equation of motion (17.18) reduces to

$$\left(\nabla^2 - (mv)^2\right) \mathbf{E} = 0, \tag{17.19}$$

where we have included a velocity of light $v = 1/\sqrt{\varepsilon\mu}$ different from c in the medium. The solution with the boundary condition $\mathbf{E}(z = 0) = \mathbf{E}_{\text{ext}}$ is given by

$$E_x(z) = E_{\text{ext}}\, e^{-\frac{z}{\lambda_{\text{el}}}},$$
$$\lambda_{\text{el}} = \frac{1}{vm}. \tag{17.20}$$

Electric fields are expelled from a superinsulator in dual analogy to magnetic fields in superconductors, as shown in Fig. 17.1: this is the *electric Meissner effect*. The length λ_{el} is the corresponding *electric penetration depth*. As magnetic fields in type II superconductors, electric fields can penetrate a superinsulator in form of electric flux tubes with tension σ, which are the electric analogues of Abrikosov vortices (for a review see [5]). First penetration happens when the applied energy $2eV_{\text{ext}}$ (of Cooper pairs in this case) equals the energy of a flux tube reaching all the way through

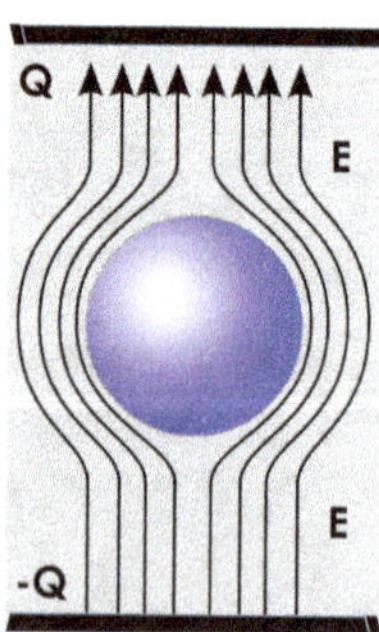

Fig. 17.1. The electric Meissner effect: electric fields are expelled from a superinsulator by dissipationless circular magnetic monopole currents.

the sample length L, $2eV_{c1} = \sigma L$. This defines the lower critical electric field,

$$E_{c1} = \frac{\sigma}{2e}.$$ (17.21)

Of course there is also an upper critical field E_{c2} corresponding to the value above which the magnetic monopole condensate is destroyed and the superinsulator transforms into a normal insulator.

Superinsulators are not the only material that expels electric fields. Metals do too. So, what is the difference between a metal and a superinsulator? If we insert a dielectric between the plates of a capacitor with accumulated charges $\pm Q$ (after we remove the battery), the applied electric field due to these charges is screened by the internal field created by the polarization of the molecules in the dielectric, resulting in an effective voltage $V_{\text{diel.}} = V_{\text{air}}/\varepsilon$, so that the capacitance $C_{\text{diel.}} = Q/V_{\text{diel.}} = \varepsilon Q/V_{\text{air}} = \varepsilon C_{\text{air}}$ is increased with respect to the situation with no dielectric. The quantity $\varepsilon > 1$ is called the relative dielectric permittivity (or dielectric constant) of the medium. Metals are perfect dielectrics in the sense that the polarization is maximal, the entire external electric field is screened and the total field within the metal vanishes. Hence, a metal has effectively $\varepsilon = \infty$. The free internal charges of the metal organize themselves so that uniform charge distributions of total charge $\pm Q$ are accumulated on its two surfaces in correspondence to the capacitor plates. The electric field lines of the capacitor plates are thus "drawn in" the metal, where they stop at the surface.

In superinsulators, the perfect screening is not due to surface charges but, rather, to dissipationless magnetic monopole currents that create a screening electric field. To show how this comes about we recall that the fundamental field $B_{\mu\nu}$ plays the role of the electromagnetic tensor field in the monopole condensate. Combining this with (16.25) we obtain the result that the monopole current in the condensate is given by

$$j_g^\mu = -\frac{1}{4\pi}\epsilon^{\mu\nu\alpha\beta}\partial_\nu B_{\alpha\beta}.$$ (17.22)

Using this we can rewrite the "electric" components of the equation of motion (17.14) as

$$\frac{2\pi}{e}\nabla \wedge \mathbf{j}_g = \frac{1}{\lambda_{\text{el}}^2}\mathbf{E},$$ (17.23)

which is the "electric London equation". Combining this with the dual Ampère law (in absence of time-varying magnetic fields)

$$\nabla \wedge \mathbf{E} = -\frac{2\pi}{e}\, \mathbf{j}_g, \tag{17.24}$$

we obtain the screening of external electric fields by internal electric fields generated by dissipationless monopole currents in the condensate. As a consequence, electric field lines are repelled, instead of drawn in, as shown in Fig. 17.1. Since the voltage between the capacitor plates is the line integral over the electric field lines and these are much longer than without the superinsulator, having to deviate to avoid it, the voltage in presence of the superinsulator is larger than without and thus $C_{\mathrm{SI}} = \varepsilon C_{\mathrm{air}}$ with $\varepsilon < 1$. A dielectric permittivity smaller than 1 is what we call *antiscreening*. A very large bulk of superinsulator is a material with $\varepsilon = 0$, the exact opposite of a metal.

Filling the upper half-space with a superinsulator is a *Gendankenexperiment*. In practice, only large slabs are possible. However, for the moment, also these are impractical, only thin superinsulating films are available. Using films between capacitor plates to detect the Meissner effect and antiscreening with an electric field along the film, however, is bound to be unsuccessful. The electric field lines will avoid the film by deviating along the thin direction and the resulting effect will typically be too small to detect. Another possibility is to apply a perpendicular electric field $\mathbf{E}_{\mathrm{ext}}$ to an extremely thin film of thickness d such that $d \ll \lambda_{\mathrm{el}}$. In this case we expect no screening in the perpendicular direction and the z-component of the electric field E^3 within the film to be a z-independent function: $E^3(r, \theta, z) = E^3(r)$, where we have assumed radial symmetry of the capacitor plates above and below the film. If these are circular, with a radius R matching the radius of a circular film between them, the applied electric flux is $\Phi = \pi R^2 E_{\mathrm{ext}}$. In this geometric setting, Eq. (17.19) becomes

$$\left(\partial_r^2 + \frac{1}{r}\partial_r - \frac{1}{\lambda_{\mathrm{el}}^2}\right) E^3 = 0. \tag{17.25}$$

The solution of this equation for the electric field inside the film is

$$E^3(r) = \frac{\Phi}{2\pi\lambda_{\mathrm{el}}^2} K_0\left(\frac{r}{\lambda_{\mathrm{el}}}\right) = \frac{1}{2} E_{\mathrm{ext}}\left(\frac{R}{\lambda_{\mathrm{el}}}\right)^2 K_0\left(\frac{r}{\lambda_{\mathrm{el}}}\right), \tag{17.26}$$

where K_0 is the zeroth-order modified Bessel function of the second kind and the overall factor is determined by the condition that the total flux through the film is the same as the applied one. The response of the superinsulator to a perpendicular applied electric field is thus to squeeze the corresponding flux onto a film area of radius $\lambda_{\rm el}$.

In the rest of this chapter we shall focus on the properties of closed confining strings. Specifically, we shall ask the question if a continuum theory for fundamental strings can be obtained by removing the UV cutoff, or, equivalently, if the strong coupling limit of compact QED is a string theory. As we have seen, confining strings have an induced action governed by an antisymmetric tensor field. In Euclidean space, it is defined by

$$e^{-S_{\rm CS}} = \frac{1}{Z(B_{\mu\nu})} \int \mathcal{D}B_{\mu\nu} \exp\left[-S(B_{\mu\nu}) + i \int d^4x \ B_{\mu\nu}T_{\mu\nu}\right], \qquad (17.27)$$

where

$$T_{\mu\nu}(\mathbf{x}) = \frac{1}{2} \int d^2\sigma \ X_{\mu\nu}(\sigma) \ \delta^4(\mathbf{x} - \mathbf{x}(\sigma)),$$

$$X_{\mu\nu} = \epsilon^{ab} \frac{\partial x_\mu}{\partial \sigma^a} \frac{\partial x_\nu}{\partial \sigma^b}, \qquad (17.28)$$

with $\mathbf{x}(\sigma)$ parametrizing the world-sheet and $S(B_{\mu\nu})$ given by (17.10).

To proceed we integrate out the antisymmetric tensor $B_{\mu\nu}$, to obtain

$$S_{\rm CS} = \int d^4x \left(T_{\mu\nu} \frac{e^2\Lambda^2}{m^2 - \nabla^2} T_{\mu\nu} + 2e^2 \partial_\nu T_{\mu\nu} \frac{1}{m^2 - \nabla^2} \partial_\alpha T_{\mu\alpha}\right). \qquad (17.29)$$

In the long-distance limit $x \gg \lambda_{\rm el}$ we can use the representation (17.28) in (17.29) and perform a derivative expansion in inverse powers of m. We start by noting that the Yukawa Green's function in (17.29) is given by

$$G(x) = \frac{1}{m^2 - \nabla^2} \delta^4(x) = \frac{m^2}{4\pi^2} \frac{1}{mr} K_1(mr), \qquad (17.30)$$

with $r = |x|$ and K_1 a modified Bessel function of the second kind. In computing (17.29) using the representation (17.28) we encounter the expression $G\left(\sqrt{g_{ab}(\sigma)\epsilon^a\epsilon^b}\right)$ with

$$g_{ab} = \frac{\partial x_\mu}{\partial \sigma^a} \frac{\partial x_\mu}{\partial \sigma^b}, \qquad (17.31)$$

the induced metric on the world-sheet. The derivative expansion of this Green's function on the world-sheet is obtained by expanding $\int d^2\epsilon \ G\left(\sqrt{g_{ab}\epsilon^a\epsilon^b}\right) f(\epsilon)$ in powers of $1/m$ for any test function $f(\epsilon)$. Naturally, the coefficients of this expansion may diverge and must then be regularized by the ultraviolet cut-off Λ_0. We obtain

$$G\left(\sqrt{g_{ab}\epsilon^a\epsilon^b}\right) = \frac{1}{2\pi}g^{-1/2}K_0\left(\frac{m}{\Lambda_0}\right)\delta^2(\epsilon) + \frac{1}{4\pi m^2}g^{-1/2}g^{ab}\partial_a\partial_b\delta^2(\epsilon) + \dots,$$

$$g = \det g_{ab} = \frac{1}{2}X_{\mu\nu}X_{\mu\nu}. \tag{17.32}$$

Inserting this expression in (17.29) we obtain the desired derivative expansion of the confining string action S_{CS},

$$S_{\text{CS}} = \frac{e^2\Lambda^2}{4\pi}K_0\left(\frac{m}{\Lambda_0}\right)\int d^2\sigma\,\sqrt{g} - \frac{\Lambda^2}{16\pi m^2}\int d^2\sigma\,\sqrt{g}g^{ab}\partial_a t_{\mu\nu}\partial_b t_{\mu\nu}$$

$$+ \frac{\Lambda^2}{16\pi m^2}\int d^2\sigma\,\sqrt{g}R + \dots, \tag{17.33}$$

where

$$t_{\mu\nu} \equiv g^{-1/2}X_{\mu\nu}, \tag{17.34}$$

and R is the scalar curvature of the world-sheet. The first term in the expansion (17.33) is the dominant Nambu–Goto term (for a review see [162]), proportional to the area of the world-surface, the equivalent of the mass term for a particle. The second and third term represent the extrinsic and intrinsic curvature terms, respectively, and describe the geometric characteristics of the electric flux tube [163]. Note that the extrinsic curvature term enters this derivative expansion with a negative coefficient. All higher order terms are suppressed by negative powers of m and are thus infrared irrelevant.

Using expression (17.16) for the mass m we can further simplify the action to

$$S_{\text{CS}} = \frac{e^2\Lambda^2}{4\pi}K_0\left(\frac{\sqrt{z}}{4e}\right)\int d^2\sigma\,\sqrt{g} - \frac{e^2}{16\pi}\int d^2\sigma\,\sqrt{g}g^{ab}\partial_a t_{\mu\nu}\partial_b t_{\mu\nu}$$

$$+ \frac{e^2}{16\pi}\int d^2\sigma\,\sqrt{g}R. \tag{17.35}$$

As stressed above, this derivative expansion is valid on scales much larger than $1/m$, with $m = \Lambda/e$. Moreover this result is valid only in the strong coupling limit $e > e_{\text{cr}}$, above the critical coupling for monopole condensation. The continuum limit we would like to take is thus $\Lambda_0 \to \infty$ with $e \to$ const. or $e \to \infty$ such that $m \to \infty$. However, in all these cases the string tension diverges, since $\lim_{e\to\infty} K_0(\sqrt{z}/4e) = \lim_{e\to\infty} K_0(\text{const.}/e) = \infty$. Compact QED in 3D is a cutoff theory and monopole condensation is not a continuous phase transition [142].

This result, however, is completely modified by the presence of a θ-term (see Chapter 16). In this case, the low-energy electromagnetic response of a superinsulator is changed from (17.10) to

$$S_\theta\left(B_{\mu\nu}\right) = \int d^4x \left(\frac{1}{12\Lambda^2} H_{\mu\nu\alpha} H_{\mu\nu\alpha} + \frac{1}{4e^2} B_{\mu\nu} B_{\mu\nu} + \frac{i\theta}{32\pi^2} B_{\mu\nu} \epsilon_{\mu\nu\alpha\beta} B_{\alpha\beta} \right).$$

$$(17.36)$$

Correspondingly the mass m is changed to m_θ given by

$$m_\theta = \frac{e\Lambda}{4\pi} \sqrt{\left(\frac{4\pi}{e^2}\right)^2 + t^2},$$

$$t = \frac{\theta}{\pi}.$$

$$(17.37)$$

We will call such superinsulators with a non-vanishing θ-term *oblique superinsulators* for reasons to become clear in a later chapter. For strong coupling, the modified mass reduces to

$$m_\theta \simeq \frac{e\Lambda t}{4\pi}, \qquad \text{for } e \gg 1, \tag{17.38}$$

and, repeating the steps leading to (17.33), we obtain the modified derivative expansion

$$S_{\mathrm{CS}}^\theta = \frac{\Lambda^2}{4\pi} K_0 \left(\frac{et}{4\pi}\right) \int d^2\sigma \ \sqrt{g} - \frac{1}{16\pi} \left(\frac{4\pi}{et}\right)^2 \int d^2\sigma \ \sqrt{g} g^{ab} \partial_a t_{\mu\nu} \partial_b t_{\mu\nu}$$

$$+ \frac{1}{16\pi} \left(\frac{4\pi}{et}\right)^2 \int d^2\sigma \ \sqrt{g} R - i \frac{\pi t}{\left(\frac{4\pi}{e^2}\right)^2 + t^2} \nu + \dots, \tag{17.39}$$

of the confining string action. Two changes appear as a consequence of the θ-term: first, the coefficients of the Nambu–Goto and curvature terms are modified; secondly, a new marginal term appears. This is a topological term involving

$$\nu = \frac{1}{4\pi} \int d^2\sigma \ \sqrt{g} \ \epsilon^{\mu\nu\alpha\beta} g^{ab} \partial_a t_{\mu\nu} \partial_b t_{\alpha\beta}, \tag{17.40}$$

the (signed) *self-intersection number* of the world-sheet in 4D Euclidean space.

For particles in 1D, the spin can be accounted for by a factor $\exp(i\theta w)$ in the partition function, where w is the signed self-intersection number of the world-lines on the Euclidean plane, called the Whitney index [164]. Bosons are realized for $\theta = 0 \bmod 2\pi$, fermions for $\theta = \pi \bmod 2\pi$. Fermions, therefore, carry a factor $(-1)^w$ in the partition function. This led Polyakov to call the factor $\exp(i\pi/t)\nu$ the "spin factor of strings" and to propose the presence of this factor with $t = 1$, corresponding to "fermionic strings", as a generic stabilization mechanism for strings [19, 163]. That this mechanism indeed works has then been proved in [159]. To see this, let us consider the large-x asymptotic behaviour

$$K_0(x) \simeq \frac{e^{-x}}{\sqrt{x}}. \tag{17.41}$$

As a consequence we can now take the simultaneous limits $\Lambda_0 \to \infty$ and $e \to \infty$ so that

$$\frac{\Lambda^2}{4\pi} \frac{e^{\frac{-et}{4\pi}}}{\sqrt{\frac{et}{4\pi}}} = T, \tag{17.42}$$

defines the physical string tension T. In this limit the curvature terms vanish and the confining string action reduces to

$$S^\theta_{\mathrm{CS}} = T \int d^2\sigma \sqrt{g} - i\frac{\pi}{t}\nu. \tag{17.43}$$

This is the action of the spinning Nambu–Goto string. In presence of a θ-term, the monopole condensation becomes a continuous phase transition separating a weak-coupling gauge theory from a strong-coupling string theory.

The expression (17.42) for the string tension can be rewritten as

$$\ln \frac{\Lambda^2}{4\pi T} = \frac{et}{4\pi} + \frac{1}{2} \ln \frac{et}{4\pi}. \tag{17.44}$$

For $e \gg 1$ the logarithm can be neglected with respect to the linear term and thus

$$e = \frac{4\pi}{t} \ln \frac{\Lambda_0^2}{4\pi T}. \tag{17.45}$$

As usual we can express the UV cutoff in terms of the physical string tension T and a reference scale μ as $\Lambda_0 = \sqrt{4\pi T}/\mu$ so that $\Lambda_0 \to \infty$ corresponds to the infrared limit $\mu \to 0$. This gives us the running coupling constant

$$e(\mu) = \frac{4\pi}{t} \ln \frac{T}{\mu^2}. \tag{17.46}$$

In the infrared limit $\mu \to 0$ we have $e \to \infty$. At a finite scale, this equation determines the perturbative corrections, like the curvature terms, to the free string with topological term.

Chapter 18

Electric pions and asymptotic freedom

In this chapter we will focus again on 2D superinsulating films. We have shown that the electromagnetic response of such materials is compact QED. Compact QED in 2D, however, does not have a phase transition, it is always in a confining phase, since monopole instantons are always in a plasma phase. Things are different, however, when compact QED is coupled to matter. Dynamical matter can induce a deconfinement phase transition [165].

In our case, compact QED is the electromagnetic response only deep in the superinsulating phase. Near the superconductor-to-insulator transition (SIT) the interaction of two types of dynamical matter, charges and vortices, becomes important. In Chapter 10 we have derived that, near the transition to a Bose metal, superinsulators are characterized by two dimensionless coupling constants, the conductance g, the ratio of the energy scales of the two types of dynamical matter, and the entanglement parameter z, which determines the strength of the charge confining interaction: $z = 0$ for total confinement, $z > 0$ for partial one. The SIT corresponds to an IR Berezinskii–Kosterlitz–Thouless [14] fixed point ($g_{\mathrm{cr}}, z = 0$). The BKT renormalization flow towards short UV scales implies a decreasing g and an increasing z. The confining interaction decreases when flowing towards short scales. This is the phenomenon of *asymptotic freedom*: the interaction becomes weak at short scales. Asymptotic freedom is typically associated with non-Abelian gauge theories, where it characterizes their UV fixed point [4]. There are, however, many other field theories that display this phenomenon, most notably the non-linear sigma model, the Gross–Neveu model and the sine–Gordon model in 2D, which is equivalent to the XY model [166]. In our case, asymptotic freedom is *not* associated with compact QED but, rather, with the sine–Gordon model. It describes the behaviour at the IR fixed point, not the UV one and, as such it should be called actually *asymptotic safety* since there is no guarantee of a UV Gaussian fixed point [166]. We will however stick with the more familiar concept.

Of course, this BKT transition should be reflected also in the electromagnetic response. As we have seen, the divergence of ε near the SIT (at fixed screening length) makes compact QED more and more non-relativistic there. Only the electric fields matter and, as a consequence, the monopole instanton interaction, from $1/r$ deep in the superinsulation region, becomes logarithmic near the SIT. Monopole instantons undergo themselves a BKT transition that matches the deconfinement of charges.

In superinsulators charges are confined by strings, made of squeezed electric flux tubes. Strings of typical width $\lambda_{\rm el}$ and typical length $d_s = 1/\sqrt{\sigma}$ (see Sec. 5.3) are the only excitations for applied voltages below the lower critical field $E_{c1} = \sigma/2e$, the Meissner state of the superinsulator. Such a string, with a Cooper pair and a Cooper hole at its endpoints, is the electric equivalent of a strong interaction *pion*. Of course, the electric interaction is much weaker than the strong interaction. Therefore, we expect the size of such an electric pion to be much larger than a real pion. This makes it possible to actually directly observe the interior of such a bound state, which is out of reach for real pions, where only indirect evidence at collider experiments is available. In particular, asymptotic freedom implies that the string becomes "loose" at short distances and that we recover the logarithmic Coulomb interaction below the screening length $\lambda_{\rm el}$. If the electric pion size is sufficiently large we could access the asymptotically free regime by simply experimenting on ever smaller samples.

The interaction potential between Cooper pairs in the superinsulator is

$$U(r) = \sigma(T)r - \frac{\pi}{24r} + a\left[\ln\left(\frac{\lambda_{\rm el}}{r_0}\right) - K_0\left(\frac{r}{\lambda_{\rm el}}\right)\right], \qquad (18.1)$$

where the first term is the linear string potential, the second term is a universal long-distance string correction called the Lüscher term [167], and the third term is the screened 2D Coulomb potential (K_0 is the MacDonald function that we have encountered many times) that reduces to $a\ln(r/r_0)$ at $r \ll \lambda_{\rm el}$ and decays exponentially at $r \gg \lambda_{\rm el}$. Here r_0 is a UV cutoff of the order of the coherence length ξ. Since the Lüscher term is a finite-size correction for $r \to \infty$ we shall henceforth neglect it. The potential is normalized so that $U(r_0) \simeq 0$. The strength of the Coulomb potential is

$$a = \frac{4e^2}{2\pi\varepsilon d}.$$ (18.2)

When samples are very big, with their dimension L such that $L \gg d_s$, charges are confined and we expect the usual hyperactivated behaviour of the resistance as a function of temperature of superinsulators. For smaller samples, however, things are different. The most interesting situation occurs for samples with dimensions in the range $\lambda_{\mathrm{el}} < L < d_s$. In this case, Cooper pairs sufficiently far apart feel neither the string tension, since the string is loose on these scales, nor the Coulomb interaction, which is screened on the scale λ_{el}. We expect thus to observe a transition from hyperactivated resistance behaviour to a metallic saturation at the lowest temperatures when the sample size is decreased. This is exactly what has been observed in a NbTiN superinsulating film by varying the bridge length on which the external voltage is applied [22], as shown in Fig. 18.1. For large bridge lengths the film displays hyperactivated resistances, for the smallest bridge length 0.2 mm, however, we see metallic saturated behaviour at low temperatures. The crossover from hyperactivation to metallic behaviour should take place around a bridge length $L \approx d_s$. The typical string size can be estimated from experimental data as follows. The energy $k_B T_{\mathrm{dec}}$ is the energy necessary to break up the string by raising the temperature. So it is a measure of $\sqrt{\sigma}$ and, therefore,

$$d_s \approx \frac{\hbar v}{k_B T_{\mathrm{dec}}},$$ (18.3)

where we have reinstated physical units with $v = (1/\sqrt{\varepsilon})c$. Using the experimentally determined deconfinement temperature $T_{\mathrm{dec}} \approx 400$ mK and the known dielectric constant of NbTiN near the SIT [15], $\epsilon \approx 800$, one can obtain an estimate $d_s \approx 0.13$ mm in excellent quantitative agreement with the observation of the metallic crossover.

We now focus on the current–voltage characteristics of the superinsulating state. To determine the $I(V)$ response we calculate the current as $I \propto 2en_{\mathrm{f}}V$, where n_{f} is the equilibrium density of free charges. For an external electric field $E_{\mathrm{ext}} < E_{\mathrm{c1}} = \sigma/2e$ the maximum of the potential lies always at the distance L corresponding to the sample size and, thus, the current is simply proportional to the number of charges activated over the

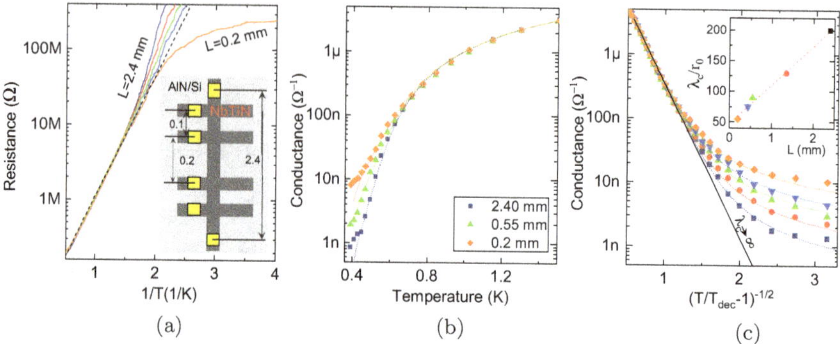

Fig. 18.1. Sheet resistance of a NbTiN superinsulating film as a function of bridge length (effective sample size). (a) Logarithmic plot of sheet resistance $R_\square$ vs. inverse temperature $1/T$ for bridges of various length L. The dashed straight line shows the Arrhenius behaviour $R \propto \exp(E/T)$. Inset: experimental setup. The Si substrate with AlN buffer layer is shown with light gray and the Hall bridge of NbTiN is dark grey. The square gold contacts are given in yellow. All lateral sizes are given in millimetres. (b) Same data as in (a) but replotted in terms of the conductance $G = 1/R_\square$ vs. T in log-line scale. A few curves are omitted to avoid overcrowding. The dotted lines are fits using a two-dimensional Coulomb gas model that generalizes the Berezinskii–Kosterlitz–Thouless (BKT) formula for the conductance $G \propto \exp[-(T/T_{\mathrm{dec}} - 1)^{1/2}]$ by incorporating a self-consistent solution of the effects of electrostatic screening, where the screening length λ_{c} and T_{dec} enter as fitting parameters. For all bridges the deconfinement temperature is $T_{\mathrm{dec}} \approx 400$ mK. (c) Same data as in (b) but for temperature renormalized as $(T/T_{\mathrm{dec}} - 1)^{1/2}$. The solid line corresponds to the case of an infinite electrostatic screening length $\lambda_{\mathrm{c}} \to \infty$. From M. C. Diamantini, L. Gammaitoni, C. Strunk, S. V. Postolova, A. Yu. Mironov, C. A. Trugenberger and V. M. Vinokur, Direct probe of the interior of an electric pion in a Cooper pair superinsulator, *Nat. Comm. Phys.* **3**, 142 (2020). Creative Commons Attribution 4.0.

barrier $(\sigma - 2eE_{\mathrm{ext}})L$,

$$I \propto V \exp\left(-\frac{\sigma(T)L}{k_{\mathrm{B}}T}\right), \quad V < V_{\mathrm{c1}} = \frac{\sigma(T)L}{2e}, \tag{18.4}$$

which, in the thermodynamic limit $L \to \infty$, implies an infinite resistance.

For $E_{\mathrm{c1}} < E_{\mathrm{ext}}$, instead, the potential has a maximum at $r = r^*$ determined by the equation $K_1\left(r^*/\lambda_{\mathrm{el}}\right) = (2eE_{\mathrm{ext}} - \sigma)\,\lambda_{\mathrm{el}}/a$, with K_1 a modified Bessel function of the second kind. Two distinct regimes become possible. The first is realized in small samples such that

$$K_1\left(L/\lambda_{\mathrm{el}}\right) > \frac{4\lambda_{\mathrm{el}}}{d_s}\frac{\Delta V}{V_{\mathrm{c1}}}, \tag{18.5}$$

where $\Delta V = V - V_{c1}$, and we have used the estimate (18.3), $d_s = \hbar v / k_B T_{\text{dec}}$. Then the potential for $\lambda_{\text{el}} < r < L$ is essentially flat. This is the asymptotically free regime discussed above, where charges effectively do not interact, and we expect thus a metallic saturation of the resistance at the lowest temperatures. The ratio $d_s/\lambda_{\text{el}} > 1$ but not typically extremely large [16]. Also, the function $K_1(x) \simeq \exp(-x)/\sqrt{x}$ at $x \gg 1$. Thus, the typical sample size for which this metallic behavior emerges is $O(d_s)$, as already pointed out above, although it can become larger if measurements are taken just above V_{c1}.

In the limit opposite to (18.5), the total energy $U_{\text{E}_{\text{ext}}}$ of the charge–anticharge pair following from (18.1) is

$$U_{\text{E}_{\text{ext}}} = a \ln(r/r_0) - Fr, \tag{18.6}$$

where $F = 2eE_{\text{ext}} - \sigma$ is the effective force pulling the charge–anticharge pair apart. The saddle point, r^*, of this potential, controlling the activated current, is $r^* = a/F$, so that the energy barrier is $U^* \equiv U_{\text{E}_{\text{ext}}}(r^*) = a\left[\ln(a/Fr_0) - 1\right]$. In equilibrium, the ionization rate $\mathcal{R} \propto \exp(-U^*/k_B T)$ and the recombination rate, $\mathcal{R}_{\text{r}}$, of the $\pm$ charges, are equal. Since $\mathcal{R}_{\text{r}} \propto n_+ n_- = n_{\text{f}}^2$, then, with logarithmic accuracy, $n_{\text{f}} = \sqrt{\mathcal{R}} \propto \exp(-U^*/2k_B T) \propto (Fr_0/a)^{a/(2k_B T)}$ [168]. At this point we can use the previously derived formula (9.19) for the deconfinement temperature. For the NbTiN film used in the described experiment, the combination $\varepsilon \lambda_{\text{el}}/\ell = O(100)$ so that

$$k_B T_{\text{dec}} \approx 8a. \tag{18.7}$$

This gives the $I(V)$ curve

$$I \propto (V - \sigma L/2e)^{1+T_{\text{dec}}/16T},$$
$$V_{c1} < V < V_{c2} \equiv \frac{L}{2e}\left[\sigma(T) + \frac{k_B T_{\text{dec}}}{2.718 r_0}\right], \tag{18.8}$$

where the upper critical voltage V_{c2} is the voltage where the energy barrier U^* vanishes, the $I(V)$ curve experiences a jump and the system switches from superinsulation into a normal insulating state. In the field interval such that $V_{c1} < V < V_{c2}$, the superinsulator is in the mixed state, where an ensemble of electric strings penetrates the system end-to-end. This is the analogue to the mixed, or Abrikosov state in superconductors.

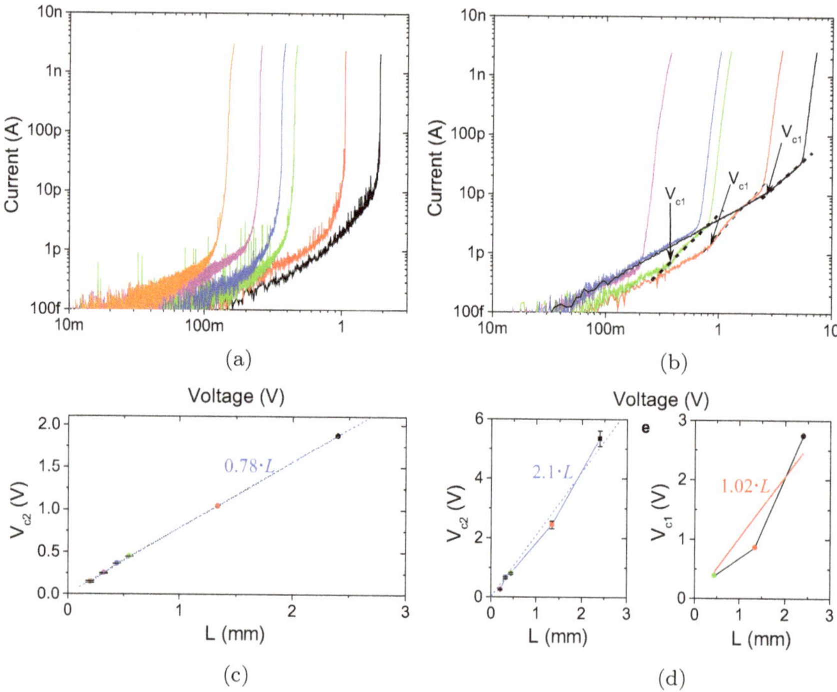

Fig. 18.2. $I(V)$ curves and threshold voltages as a function of sample size. The $I(V)$ curves of two different NbTiN samples with (a) $R_\square(T = 2\ \mathrm{K}) = 0.2\ \mathrm{M\Omega}$ and (b) $R_\square(T = 2\ \mathrm{K}) = 2\ \mathrm{M\Omega}$ measured at the same temperature $T = 50\ \mathrm{mK}$. Different colours correspond to different bridge length L (distance between electrodes). (c) Dependence of the threshold voltage V_{c2} on the bridge length for the 0.2 MΩ sample. The dashed line depicts the $V_{c2} = 0.78 \cdot L$ dependence. The error bars are defined as the width of the jump and are less than the symbol sizes. (d) Threshold voltage V_{c2} for the 2 MΩ sample; $V_{c2} = 2.1 \cdot L$ (dashed line). Error bars are given by the width of the jump. (e) The bridge length dependence of the kink voltage V_{c1}, marked by arrows in the main panel. The error bars are determined from the errors in the slopes of the $I(V)$ curves and are much less than the symbol sizes. From M. C. Diamantini, L. Gammaitoni, C. Strunk, S. V. Postolova, A. Yu. Mironov, C. A. Trugenberger and V. M. Vinokur, Direct probe of the interior of an electric pion in a Cooper pair superinsulator, *Nat. Comm. Phys.* **3**, 142 (2020). Creative Commons Attribution 4.0.

Figure 18.2 presents the threshold behaviours of the $I(V)$ curves corresponding to bridges of different length for two samples having the resistances $R_\square = 0.2$ and 2 MΩ, respectively at $T = 2$ K. The current jumps span a range of a few orders of magnitude and become less sharp upon shortening the distance between electrodes. The threshold voltage V_{c2} exhibits a linear dependence upon L, see Fig. 18.2(c) and (d), implying that the threshold electric field $E_{c2} = V_{c2}/L$ is independent of the distance between electrodes in full accordance with (18.8). For the low-resistance sample, shown in Fig. 18.2(a), the string tension can be estimated from $d_s \simeq 0.13$ mm as $\sigma \simeq 4.15 \times 10^{-21}$ J/m $= 26$ meV/m, which leads to a contribution $\sigma/2e = 1.3 \times 10^{-5}$ V/mm to the slope. This is negligible with respect to the Coulomb contribution (second term in (18.8)), which amounts to a predicted slope of 0.63 V/mm, again in remarkable quantitative agreement with the measured slope of 0.78 V/mm. The $I(V)$ characteristics below the threshold, measured at 50 mK, are linear, which agrees reasonably well with the theoretically predicted power-law exponent $1 + (T_{\mathrm{dec}}/16T) = 1 + (400 \text{ mK}/16 \times 50 \text{ mK}) = 1.5$.

The $I(V)$ characteristics of the high-resistance sample are shown in Fig. 18.2(b). The three largest bridges exhibit both kinks V_{c1} and V_{c2}, which is direct evidence for the Meissner state of superinsulators. The two smallest bridges do not resolve V_{c1} because their size is so small that strings penetrate the entire sample in the whole experimentally accessible voltage range and, thus, the Meissner state is not visible. The linear dependence $V_{c2} \propto L$ has the slope 2.1 V/mm, see Fig. 18.2(d). Three available values of V_{c1} are unfortunately not sufficient for conclusive evidence of linearity, see Fig. 18.2(e). However, if one assumes $V_{c1} \propto L$, then the mean square deviation estimate for the approximating straight line (shown in red) would give 1.02 V/mm for the slope. Making use of Eq. (18.8), one would obtain the slope $2.1 - 1.02 \approx 1.1$ characterizing the pure Coulomb contribution. From (18.8), one can then estimate the deconfinement temperature 692 mK for this sample, yielding the exponent $1 + (T_{\mathrm{dec}}/16T) = 1 + (692 \text{ mK}/16 \times 50 \text{ mK}) = 1.86$ for the $I(V)$ characteristics in the mixed state, $V_{c1} < V < V_{c2}$. Comparing this value with the measured ones for the three largest bridges, 1.6, 1.8 and 2, respectively, one can then conclude that the assumption $V_{c1} \propto L$ results in a fair quantitative agreement between the predicted and measured exponents.

Why does the kink associated with the lower critical field V_{c1} not show up in the $I(V)$ curves of the low-resistance sample too? Due to its lower resistance, this sample is closer to the SIT, where the deconfinement temperature decreases towards zero. Since measurements of both samples are made at the same temperature, this means that the low-resistance sample is closer to the deconfinement temperature, where the string tension vanishes and strings become infinitely long. As a consequence, strings in the lower-resistance sample are longer than in the high-resistance sample and penetrate the sample end-to-end for all accessible voltages, so that the lower kink V_{c1} is not observable. In order to resolve the lower kink and access the Meissner state one needs larger samples, which are sufficiently far from the SIT.

To conclude this chapter, let us stress that the observation of the kink V_{c1} is a direct experimental confirmation of superinsulation caused by electric strings due to magnetic monopole instantons.

Chapter 19

Oblique superinsulators, strong superinsulators, and high-T_c superconductivity

In the last part of the book we are going to discuss the possible relevance of magnetic monopoles for the physics of high-T_c superconductivity. Here we shall construct an effective long-distance field theory that reproduces the observed universal aspects of high-T_c superconductors [41]. In the next chapter we will then focus on a new microscopic mechanism of pairing. Contrary to the material presented in previous chapters, there is as yet no experimental evidence for these ideas, which have to be considered, for the moment, just one possible model, albeit one that seems to explain most, if not all, the present observations.

As we have discussed in previous chapters, in 2D there exists a Bose metal behaving as a fermion system even if the material constituents are charge $2e$ Cooper pairs. The reason is that this state is a bosonic topological insulator, in which low-energy excitations live on the 1D boundary, where there is no distinction between bosons and fermions. In 3D, the bosonic topological insulator, even the strong one, has no low-energy fermionic boundary excitations which do not break time-reversal symmetry [136], so this mechanism does not work anymore. But, is there any need of such a mechanism? Does a 3D Bose metal exist? The answer is yes, it is the pseudogap state of high-T_c superconductors, which displays the resistance $\propto T^2$ characteristic of Fermi liquids [169, 170] while its constituents are still predominantly charge $2e$ Cooper pairs [171].

High-T_c superconductors are still the biggest mystery of condensed matter physics, even almost 40 years after their discovery. Tons of ink have gone into their detailed description, here we refer to [42, 172] for recent reviews and we focus only on the main characteristics of cuprates, those that must be reproduced by a successful long-distance field theory.

Cuprate superconductors are layered materials. Each layer is made of two CuO_2 conducting planes with various possible atoms in between, like Ca or Y. Superconductivity happens along these 2D horizontal planes and does not depend on the number of stacked layers, one layer is sufficient [173]. Given that electrons can tunnel between the atomic-distance planes, however, this single-layer structure is already 3D, albeit very anisotropic. As we shall see, this 3D aspect is crucial for superconductivity. The quantum driving parameter is the doping p, typically the concentration of holes (removed electrons), although also doping by adding electrons is considered. Undoped materials are antiferromagnetic insulators, upon doping superconductivity is induced. This is the first indication that the transition to superconductivity upon doping is a 3D superconductor-to-insulator transition (SIT) as discussed in previous chapters. The phase diagram of high-T_c cuprates is shown in Fig. 19.1. At low temperatures and very low doping, strong electron interactions localize the antiferromagnetic state into a Mott insulator [169, 170, 174]. Further doping turns the electrons itinerant and the superconducting dome forms in the $p \approx 0.05$–0.3 interval, followed by a Fermi liquid at even higher doping. The transition temperature $T_c(p)$ is defined by the contour of this dome, the highest value being typically reached for a p_{cr} around the dome centre. Values of $p < p_{cr}$ define the underdoped region, values $p > p_{cr}$ the overdoped region.

In the underdoped regime, a mysterious state, called the *pseudogap state* (PG), opens up at a temperature T^* much above T_c (see e.g. [42]). Above the temperature T^* a metallic sheet resistance linear in temperature, $R_\square \propto T$, is observed. At the same time, below a temperature $T^{**} < T^*$, a distinct switch to a pure Fermi liquid, quadratic $R_\square \propto T^2$ dependence occurs, which holds till superconducting fluctuations set in near the dome, see e.g. [169, 170]. The full switching on of the PG state at $T = T^{**}$ is accompanied by a nematic phase transition [175], where the electronic C_4 symmetry of the CuO_2 cell is broken to diagonal C_2, and by the emergence of the magnetoelectric Kerr effect [174–177], evidencing that the PG state is a distinct thermodynamic phase. Recent shot noise measurements deep in the PG region detected pairing of the charge carriers in this phase [171].

Another mystery of high-T_c superconductors concerns the overdoped region. If a sufficiently high magnetic field is applied, so that superconductivity is destroyed, a linear $R_\square \propto T$ resistance behaviour is observed

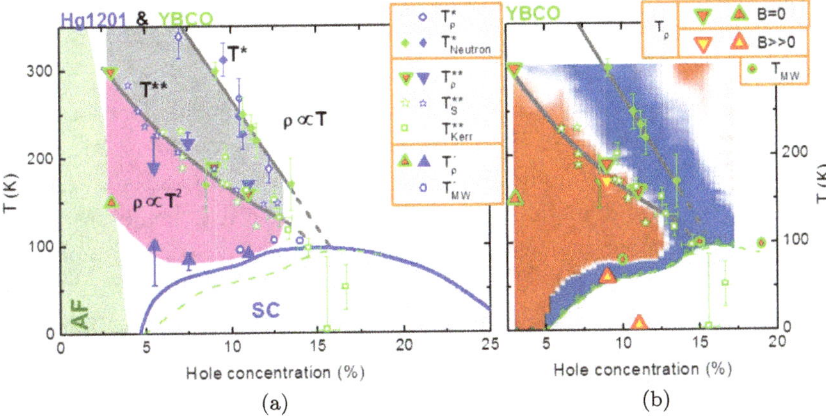

Fig. 19.1. Phase diagram of the Hg1201 and YBCO high-T_c superconductors. (a) Pseudogap temperature T^*, determined from the deviation from linear-in-T resistivity (T^*_ρ) and neutron scattering experiments (T^*_{neutron}). The gray shaded area indicates the crossover to the quadratic regime (magenta) found below T^{**}. The latter ends with the onset of SC fluctuations at T', in agreement with microwave (T'_{MW}) measurements for Hg120149 and YBCO50, or when localization effects set in (around 150 K for YBCO at $p = 0.03$). The temperatures of the TEP peak (T^{**}_S) for Hg120122 and YBCO and of the onset of the Kerr effect (T^{**}_{Kerr}) for YBCO25 track T^{**}_ρ from dc-resistivity. Blue and green symbols correspond to Hg1201 and YBCO, respectively. Green dashed line corresponds to the $T_c(p)$ of YBCO. The blue line is obtained from available data for $T_c(p)$ of Hg1201 (up to $p = 0.21$) and extended to higher doping. Gray lines for $T^*(p)$ and $T^{**}(p)$ are guides to the eye. The antiferromagnetic (AF) phase is schematically indicated by the green shaded area. (b) The underlying T^2 (red contour) regime of YBCO is effectively captured by a map of the resistivity curvature. The quadratic resistive behaviour is also apparent after applying a high c-axis magnetic field of $B \simeq 50$ T. For $p = 0.11$, the field was sufficiently high to completely suppress superconductivity and reveal the approximately quadratic resistive behaviour to low temperatures. From N. Barišić *et al.*, Universal sheet resistance and revised phase diagram of the cuprate high-temperature superconductors, *PNAS* **110**, 12235–12240 (2013), with permission by the authors and the publisher, ©PNAS (2013).

at the lowest temperatures [178]. While its generic character is established experimentally and is typically associated with a universal scattering rate, its origin remains a mystery, since phonons, typically responsible for a linear-in-T resistance, are frozen at these temperatures. A careful experimental analysis [179] reveals that the slopes of the linear terms in the low-temperature regime and in the high-temperature, inherently metallic phase, are noticeable different. Therefore, the low-temperature linear-in-T regime is clearly different from the metallic behaviour above the superconducting dome. Moreover, at sufficiently low temperatures and high magnetic field

B there is an intercept which is also linear in B [179], so that the resistivity assumes the form $\rho = aT + bB$. These two can be considered as thermal and quantum contributions, respectively.

The evolution from a Mott insulator at $p = 0$ to superconductivity upon increasing p is reminiscent of a quantum superconductor-to-insulator transition (SIT) [180] at the putative transition point p^* where the $T^*(p)$ line continued into the dome hits the p-axis. Furthermore, it has been conjectured that the superconducting dome may be a coexistence region for two quantum phases. Combined with the recent observation of an emergent granular structure in high-T_c materials [37], all these facts suggest that high-T_c cuprates realize a direct, first-order 3D SIT as we introduced in Chapter 8 and that (14.26) is the generic electromagnetic response of the phases of high-T_c materials.

This, however, cannot be completely right yet. The magnetoelectric effect has been detected in the pseudogap phase [174–177] and this is not reflected in (14.26). However, we now know what is the origin of the magnetoelectric effect, a θ-term must be included in the electromagnetic response (Chapter 16), and its main consequence is the Witten effect [156], assigning an electric charge $\theta/2\pi$ to magnetic monopoles, making them *dyons*. We shall thus modify (14.26) to

$$
\mathrm{e}^{-S_{\mathrm{HTS}}(\mathcal{A}_\mu)} = \frac{1}{Z} \sum_{\{Q_\mu\},\{M_\mu\}} \int \mathcal{D}\mathcal{A}_\mu \mathrm{e}^{-S},
$$

$$
S = \sum_x \left[\frac{1}{4f^2} \left(\mathcal{F}_{\mu\nu} - 2\pi S_{\mu\nu}\right)^2 - i\mathcal{A}_\mu \left(Q_\mu + \frac{\theta}{2\pi} M_\mu\right)\right],
\tag{19.1}
$$

where the integers Q_μ and $M_\mu = (1/2)K_{\mu\alpha\beta}S_{\alpha\beta}$ are the conserved charge and magnetic monopole currents, respectively, $K_{\mu\alpha\beta}$ is the lattice BF operator introduced in Chapter 14 and we have rescaled the effective Coulomb interaction strength $f \to \sqrt{2}f$ to match the usual normalization. As in previous chapters, we denote by $\mathcal{A}_\mu$ the dimensionless gauge fields rescaled with the granular size, $\mathcal{A}_\mu = \ell A_\mu$. We have omitted the Q_μ^2 term in (14.26) since it will be irrelevant in the limit of large granular size $\ell\Lambda \gg 1/f$ in the notation of Chapter 14 that we will consider henceforth. Having in mind Cooper pairs, we let the electric charge unit be $2e$. The Dirac quantization condition then requires the unit of magnetic charge to be π/e.

This action was first introduced in [181] to describe *oblique confinement*, a new phase of strongly interacting matter previously posited by 't Hooft [182] in which a *dyon condensate* carrying electric and magnetic charge (q, m) confines excitations carrying the orthogonal combination $(m, -q)$. We know from our analysis in Chapter 14 that, without the θ-term, a condensate of charges is a superconductor, a condensate of monopoles is a superinsulator, and the phase without any condensate is an insulator. When we switch on the θ-term, monopoles acquire charge and the resulting dyon condensate becomes the *oblique superinsulator* we introduced in Chapter 17. To derive the complete phase diagram we have to establish which condensates are favoured as we vary the parameters f and θ. To do so we will follow exactly the same route that we took in previous chapters and we integrate out the gauge field $\mathcal{A}_\mu$ to obtain the effective action for the two types of topological defects alone,

$$Z = \sum_{\{Q_\mu\}, \{M_\mu\}} e^{-S},$$

$$S = \sum_x f^2 \left(Q_\mu + \frac{\theta}{2\pi} M_\mu \right) \frac{1}{-\nabla^2} \left(Q_\mu + \frac{\theta}{2\pi} M_\mu \right)$$

$$+ \frac{\pi^2}{f^2} \sum_x \left(M_\mu \frac{1}{-\nabla^2} M_\mu + i\pi Q_\mu \frac{1}{-\nabla^2} K_{\mu\alpha\beta} M_{\alpha\beta} \right), \qquad (19.2)$$

where the integers $M_{\mu\nu} = (1/2) S_\mu \epsilon_{\mu\nu\alpha\beta} S_{\alpha\beta}$ are such that $M_\mu = \Delta_\nu M_{\mu\nu}$ (S_μ is the lattice shift operator, see Chapter 14). The last term represents the topological interaction between charges and monopoles. When the Dirac quantization condition is satisfied, it falls out of the partition function since it is always an integer multiple of 2π, as can be easily seen by representing the conserved charge current as $Q_\mu = K_{\mu\alpha\beta} X_{\alpha\beta}$, with the gauge conditions $\Delta_\alpha X_{\alpha\beta} = \Delta_\beta X_{\alpha\beta} = 0$, so that $X_{\alpha\beta}$ contains three degrees of freedom as Q_μ, and using the relation (14.14).

We now repeat the same analysis we presented in previous chapters. Keeping only self-interactions [96] in the above "energy" term (actually an action) and adding the corresponding "entropy" (actually quantum corrections) of lattice strings one can assign to a particle with electric charge Q and magnetic charge M a "free energy" proportional to its (Euclidean) world-line length N,

$$F = \left[f^2 G(0) \left(Q + \frac{\theta}{2\pi} M \right)^2 + \frac{\pi^2}{f^2} G(0) M^2 - \mu \right] N, \qquad (19.3)$$

where $G(0)$ is the value of the lattice Coulomb potential at coinciding points and $\mu \approx \ln 7$, since at each step the non-backtracking string has to choose among 7 possible directions to continue. If the factor in the brackets is positive, the free energy is minimized for $N = 0$. If, instead, it is negative, the free energy is minimized by $N = \infty$. This means that particles with quantum numbers resulting in a positive factor in the brackets are suppressed and exist only as short-lived fluctuations, while particles with quantum numbers for which the factor in brackets is negative form Bose condensates. If both Bose condensates are possible, the one with the lowest free energy is stable, but the phases may coexist. The condensation condition

$$\eta \frac{f^2}{\pi} \left(Q + \frac{\theta}{2\pi} M \right)^2 + \eta \frac{\pi}{f^2} M^2 < 1, \tag{19.4}$$

where $\eta = \pi G(0)/\mu$, describes the interior of a *tilted ellipse* with the semi-axes $r_Q = \sqrt{\pi/f^2}\sqrt{1/\eta}$ and $r_M = \sqrt{f^2/\pi}\sqrt{1/\eta}$ on an integer lattice of electric and magnetic charges. The quantity $1/\sqrt{\eta}$ defines the overall scale of the ellipse, while the coupling f^2/π is the ratio between its semiaxes. Varying the coupling f and the angle θ leads to complex phase structures consisting of alternating sequences of three possible phases [181], a superconducting phase consisting of a pure charge condensate ($M = 0$), a Coulomb phase with no condensate, and an oblique confinement phase, where condensed particles are dyons carrying both electric and magnetic charges.

In the original analysis of this model [181], the domain of θ was restricted to $\theta \in [0, 2\pi[$ because of the periodicity of the charge spectrum under shifts $\theta \to \theta + 2\pi$. This periodicity is evident in Eqs. (19.3) and (19.4), since the shift $\theta \to \theta + 2\pi$ can be compensated by the corresponding shift $Q \to Q - M$. The important point is that, as a consequence of this restriction, the superconducting and oblique confinement phases can never be adjacent, but are always separated by an insulating phase. However, while the charge of dyonic composites is indeed periodic under shifts $\theta \to \theta + 2\pi$, their *statistics* is not, if charges are bosons, as has been pointed out in [183]. The statistics of a composite of Q bosonic electric charges and M magnetic monopoles with $\theta = 0$ is $(-1)^{QM}$, where $+1$ corresponds to bosons and -1 to fermions, as we derived in Chapter 16. If we turn on the angle θ, the total electric charge changes to $Q_{\text{tot}} = Q + M\theta/2\pi$ but the statistics remains unchanged [153]. If we compensate the shift $\theta \to \theta + 2\pi$ by the corresponding transformation $Q \to Q - M$, to leave the total charge unchanged, we necessarily change the statistics of the composite by the factor $(-1)^{M^2}$, which is not necessarily unity. Only the shift $\theta \to \theta + 4\pi$, accompanied by $Q \to Q - 2M$, yields the same charge *and* statistics. For bosonic charges, thus the correct periodicity that leaves both the charge spectrum and the statistics unchanged is $\theta \to \theta + 4\pi$. Had charges been fermions, then the correct periodicity, indeed, would have been $\theta \to \theta + 2\pi$, but fermions do not Bose condense. This statistical Witten effect [183] is the same reason for which strong bosonic topological insulators are also periodic only under shifts $\theta \to \theta + 4\pi$, contrary to fermionic ones [136]. Correspondingly, we shall call oblique superinsulators with $\theta = 2\pi$ *strong superinsulators*.

The major consequence of the periodicity of θ under shifts 4π, instead of 2π, is that the model for $\theta = 2\pi$ is different from the same model at $\theta = 0$. For $\theta = 2\pi$, the (Q, M) lattice is not tilted, but all $M = \text{const.} \neq 0$ lines are shifted in their Q-values with respect to the x-axis $M = 0$, as shown in Fig. 19.2. Decreasing the coupling π/f^2 from high values, corresponding to a weak Coulomb interaction, to small ones, corresponding to a strong Coulomb interaction, means shifting the ellipse from a shape elongated along the x-axis, corresponding to a superconductor made of a pure charge condensate, to one elongated along the y-axis, corresponding to an oblique superinsulator made of a dyon condensate, passing through a circle at intermediate values. For $\eta > 1$, this circle forming mid-way contains

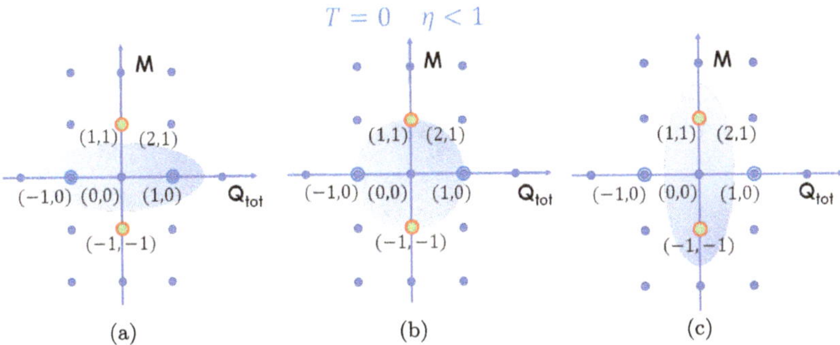

Fig. 19.2.   The quantum phases of the model for $\theta = 2\pi$. The x-axis shows the total charges $Q_{\text{tot}} = Q + M$, while the y-axis lists the magnetic charges M. (a) Superconductor: only lattice points $Q_{\text{tot}} = \pm 1$ and $M = 0$ (pure electric charges) fall inside the ellipse. (b) Coexistence of superconductor and strong superinsulator at a first-order quantum phase transition: the ellipse contains both pure and dyonic electric charges. (c) Strong superinsulator: dyons with both electric charge $Q_{\text{tot}} = \pm 1$ and $M = \pm 1$ form a condensate. From M. C. Diamantini, C. A. Trugenberger and V. M. Vinokur, Topological nature of high-temperature superconductivity, *Adv. Quantum Technologies* **4**, 2000135 (2021). Creative Commons Attribution 4.0.

only the origin of the lattice, which corresponds to an insulating phase. For $\eta < 1$, however, the deformation from superconductor to oblique superinsulator processes through a circle containing both $(Q = \pm 1, M = 0)$ and $(Q = \pm 1, M = \pm 1)$ lattice points, as shown in Fig. 19.2. This describes a first-order transition at which the superconducting and the oblique confinement orders coexist. The particles in both these coexisting condensates carry unit electric charge, but for the "superconducting particles" the charge is fundamental, while "oblique confinement" carriers acquire charge due to the Witten effect. Since $G(0) = 0.155$ and $\mu \approx \ln 7$, we have $\eta = 0.25 < 1$. It is exactly this coexistence regime that is realized, without an intervening normal insulating phase.

A finite temperature T can be introduced by considering a Euclidean time of finite length $\beta = 1/T$, with periodic boundary conditions (we have reabsorbed the Boltzmann constant into the temperature). If the original field theory model is defined on a Euclidean lattice of spacing ℓ, then β is quantized in integer multiples of ℓ as $\beta = b\ell$, with b an integer. The lattice Coulomb Green's function $G(0)$ at coinciding points is given by

$$G(0) = \frac{1}{(2\pi)^4} \int_{-\pi}^{\pi} d^4k \, \frac{1}{\sum_{i=0}^{3} 4\sin\left(\frac{k^i}{2}\right)^2}. \tag{19.5}$$

At finite temperatures, k^0 is both defined on a Brillouin zone of length $2\pi/\ell$ and also quantized in units of $2\pi/\beta$. This can be achieved only by introducing integers $k \in [-b, b]$, called Matsubara frequencies, and restricting the allowed values to $k^0 = \pi k/\ell b$. Correspondingly, integrals over k^0 have to be replaced by sums over Matsubara frequencies,

$$\int_{-\pi}^{\pi} dk^0 f\left(k^0\right) \to \sum_{k=-b}^{k=b} \frac{\pi}{b} f\left(\frac{\pi k}{b}\right). \tag{19.6}$$

As a consequence, at finite temperatures $G(0)$ is modified to

$$G(0,T) = \frac{1}{(2\pi)^4} \sum_{k=-b}^{k=+b} \frac{\pi}{b} \int_{-\pi}^{\pi} \frac{dk^1 dk^2 dk^3}{4\sin\left(\frac{\pi k}{2b}\right)^2 + \sum_{i=1}^{3} 4\sin\left(\frac{k^i}{2}\right)^2}. \tag{19.7}$$

This affects primarily the parameter η which becomes a function of the temperature, $\eta(T) = \pi G(0,T)/\mu$. Introducing the scale function $S(T) = \sqrt{G(0)/G(0,T)}$ we obtain the result that, at finite temperatures, the overall scale of the ellipse (19.4) shrinks by $S(T)$, $\sqrt{1/\eta} \to S(T)\sqrt{1/\eta}$, as shown in Fig. 19.3. Thus, lattice points that were within the ellipse at $T = 0$ fall out as the temperature grows. This leads to a tricritical point at $(p_{\rm cr}, T_{\rm cr})$ with

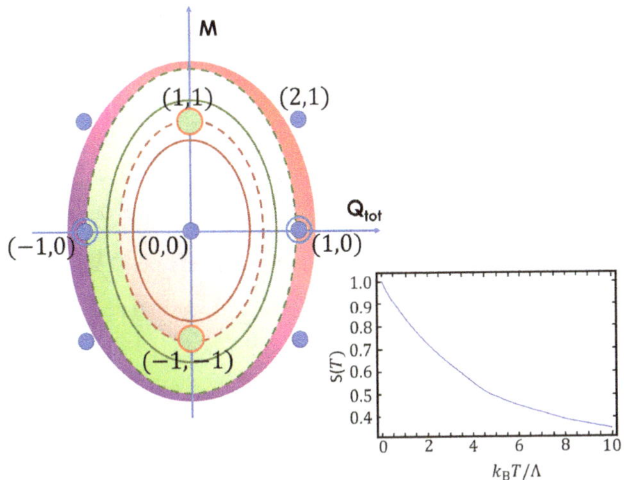

Fig. 19.3. Ellipse shrinking and deconfinement at finite temperatures for the under-doped regime $p < p_{cr}$. The purple ellipse corresponds to the coexistence of dyon and pure charge condensates. The green dashed line corresponds to the temperature at which the transition from the superconducting dome to the pseudogap phase occurs. The solid green line corresponds to temperatures at which the pseudogap phase exists. The dashed fire brick line marks the further transition from the pseudogap to a metallic state. Inset: The scaling function $S(T)$ determining the shrinking of the ellipse with increasing temperature. From M. C. Diamantini, C. A. Trugenberger and V. M. Vinokur, Topological nature of high-temperature superconductivity, *Adv. Quantum Technologies* **4**, 2000135 (2021). Creative Commons Attribution 4.0.

T_{cr} being the temperature at which the circle at fixed p_{cr} in Fig. 19.2(b) just touches the points $(Q = \pm 1, M = 0)$ and $(Q = \pm 1, M = \pm 1)$ when the temperature is varied.

The (numerically computed) phase diagram in the vicinity of the tri-critical point is shown in the inset in Fig. 19.4, where we have identified the coupling constant π/f^2 with the doping normalized to unity at the T_{cr} maximum. The long-distance field theory we have constructed is valid in a neighbourhood of this tricritical point. The value of $T_{cr} = O(100)$ K is obtained for a UV cutoff value $\Lambda \approx 2$ KeV, i.e. a frequency $\approx 1/2$ THz which is of order of the plasma frequency in cuprates. The coloured regions mark the existence of a dyon condensate with magnetic charge π/e. The super-conducting dome is identified as the coexistence regime of strong superin-sulation, the dyon condensate, and superconductivity, the pure charge condensate. In the underdoped region, the strong superinsulator is the stable state, superconductivity is realized by tunneling between metastable bubbles of pure charge condensate. The dash-dotted line hitting the $T = 0$

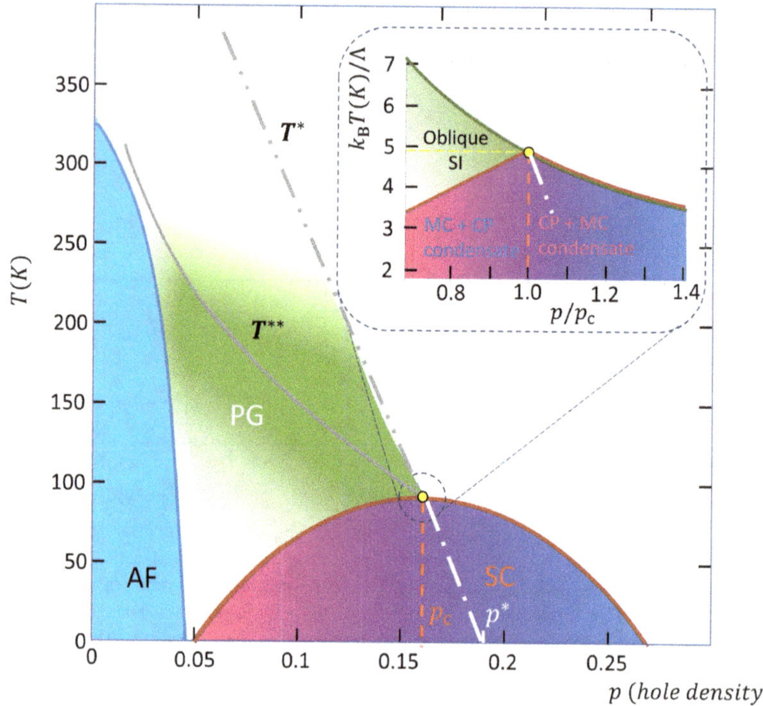

p (*hole density*)

Fig. 19.4.   The predicted phase structure of high-T_c cuprates around the tricritical point. The phase diagram depicts different phases, including the antiferromagnetic (blue), superconducting (purple to dark blue), and pseudogap (green) domains. The superconducting dome is a coexistence phase of pure Cooper pairs (CP) and charged monopoles (MC) condensates. The grey dashed-dotted line at T^* marks the onset of the pseudogap phase, detected by deviation from the linear $R_\square \propto T$ resistance dependence. The grey line T^{**} indicates the emergence of the $R_\square \propto T^2$ resistance, nematicity, and magneto-electric effects. The doping p_{cr} marks the tricritical point at the dome maximum, and p^*, the point where the $T^*(p)$ line hits the x-axis, is the position of the putative quantum Mott–insulator–superconductor phase transition at $T = 0$. Inset: the calculated phase transition lines in the vicinity of the tricritical point. From M. C. Diamantini, C. A. Trugenberger and V. M. Vinokur, Topological nature of high-temperature superconductivity, *Adv. Quantum Technologies* **4**, 2000135 (2021). Creative Commons Attribution 4.0.

axis at $p = p^*$ depicts the first-order phase transition line, which does not necessarily have to be a vertical line. When the temperature is raised the superconducting pure charge condensate ceases to exist and free charges $2e$ become short-lived excitations. Only the dyon condensate, i.e. the strong superinsulator survives: this is the pseudogap state. Note that single electrons of charge e are even more strongly suppressed in this region since they are topologically forbidden by the Dirac quantization condition, since, in presence of magnetic charges π/e they would carry an angular momentum $1/4$ [17]. In the overdoped region, $p > p_{\rm cr}$, the pure charge superconducting condensate is the stable state. Here, the boundary of the superconducting dome marks the region where dyons cease to exist. At these very high temperatures paired electrons stick together only because of the topological obstruction due to the Dirac quantization condition. As soon as the glue provided by magnetic charges is not around anymore, paired electrons will split to form a normal metal. The physics of high-T_c cuprates is thus dominated by the *tricritical point* at $p/p_{\rm c} = 1$ and $k_{\rm B}T_{\rm c}/\Lambda = 4.9$, where the first-order transition within the superconducting dome meets two continuous transitions to a normal metal (in the overdoped region) and to the pseudogap state (in the underdoped region). This universal structure is an implication of the presence of a charged magnetic monopole condensate.

We have identified the pseudogap state with a strong superinsulator, i.e. a dyon condensate of magnetic monopoles carrying both magnetic and electric charge ± 1 (in units π/e and $2e$) due to the Witten effect induced by a θ-term with angle 2π. The magnetoelectric effect is present in such a state by construction, due to the θ-term, as we explained in detail in Chapter 16. But what about the Fermi-liquid behaviour $R_\square \propto T^2$? To see how this arises let us go back to Chapter 17, where we derived the electromagnetic response of oblique superinsulators in terms of an antisymmetric tensor field. As we stressed there, the actions (17.10) and (17.37) are gauge-fixed expressions in which the massive field due to the monopole or dyon condensate "eats up" the photon in a generalized Stückelberg mechanism [160]. To make the gauge structure explicit, we separate the (Minkowski space-time) Lagrangian in terms of the massless Kalb–Ramond gauge field carrying one degree of freedom due to the monopole or dyon condensate and the original vector gauge field carrying the remaining two,

$$\mathcal{L}_\theta = \frac{1}{12\Lambda^2} H_{\mu\nu\alpha} H^{\mu\nu\alpha} - \frac{1}{4f^2} \left(B_{\mu\nu} + F_{\mu\nu}\right)\left(B^{\mu\nu} + F^{\mu\nu}\right)$$

$$+ \frac{i\theta}{32\pi^2}\left(B_{\mu\nu} + F_{\mu\nu}\right)\epsilon^{\mu\nu\alpha\beta}\left(B_{\alpha\beta} + F_{\alpha\beta}\right), \tag{19.8}$$

where vector gauge invariance under transformations $A_\mu \to A_\mu + \partial_\mu\xi$ is guaranteed, as usual, by the dependence on the field strength $F_{\mu\nu} = \partial_\mu A_\nu - \partial_\nu A_\mu$, and gauge invariance of the second kind, under transformations $B_{\mu\nu} \to B_{\mu\nu} + \partial_\mu\lambda_\nu - \partial_\nu\lambda_\mu$, is realized with the simultaneous transformation $A_\mu \to A_\mu - \lambda_\nu$. This is the Stückelberg mechanism, the Anderson–Higgs mechanism without a Higgs field or, in other words, the limit of the Anderson–Higgs mechanism in which the Higgs field becomes infinitely heavy, and predates both the Anderson and the Higgs versions. When, in superconductor physics, the phase of the order parameter appears only in the combination $(\partial_\mu \varphi - A_\mu)$ with the gauge field, one is dealing with the Stückelberg mechanism, much anterior to Anderson's realization. Usually, it is the photon that "eats up" the phase of the order parameter. The dual Stückelberg formulation here, in which the scalar is encoded in the antisymmetric tensor field, presents the opposite situation. However, it still results in three degrees of freedom with mass

$$m_\theta = \frac{f\Lambda}{4\pi}\sqrt{\left(\frac{4\pi}{f^2}\right)^2 + \left(\frac{\theta}{\pi}\right)^2}. \tag{19.9}$$

The mass (19.9) is the combination of normal and topological contributions. The dyon condensate corresponding to the pseudogap state is realized for strong coupling f. The important point is that, for $f \gg 1$, the topological contribution to the mass dominates and the mass eventually diverges as $f \to \infty$. In this limit, the bulk dynamics is frozen because only the topological contribution,

$$\mathcal{L}_{\text{top}} = \frac{\theta}{32\pi^2}\left(B_{\mu\nu} + F_{\mu\nu}\right)\epsilon^{\mu\nu\alpha\beta}\left(B_{\alpha\beta} + F_{\mu\nu}\right), \tag{19.10}$$

survives in the Lagrangian. In this limit, the strong superinsulator becomes a topological phase.

Let us now consider (19.10) on an open manifold M with the boundary ∂M. The fields $B_{\mu\nu}$ and A_μ contain transverse and longitudinal modes, the latter being encoded in the gauge field λ_μ defined by $B_{\mu\nu} = \partial_\mu \lambda_\nu - \partial_\nu \lambda_\mu$ and the phase ξ defined by $A_\mu = \partial_\mu \xi$. In the topological limit, the mass diverges and all bulk transverse modes get frozen. The longitudinal modes, however, contribute a total derivative to (19.10), as is the case for the BF model representing topological insulators [136]. When the model is defined on a bounded space, this total derivative for the longitudinal modes is all that remains and it gives rise to a boundary field theory for surface modes λ_μ and ξ, governed by the Lagrangian

$$\mathcal{L}_{\partial M} = \frac{\theta}{8\pi^2}\lambda_\mu \epsilon^{\mu\alpha\nu}\partial_\alpha \lambda_\nu + \frac{\theta}{2\pi}\lambda_\mu \Phi^\mu - \frac{M}{2}v^2\left(\partial_i \xi\right)^2 - \frac{\theta^2}{2\pi^2 M}b^2 \tag{19.11}$$

$$= \frac{\theta}{8\pi^2}\lambda_\mu \epsilon^{\mu\alpha\nu}\partial_\alpha \lambda_\nu + \frac{\theta}{2\pi}\lambda_0 \Phi^0 - \frac{\theta}{\pi}b\dot{\xi} - \frac{M}{2}v^2\left(\partial_i \xi\right)^2 - \frac{\theta^2}{2\pi^2 M}b^2, \tag{19.12}$$

where $\Phi^\mu = (1/2\pi)\epsilon^{\mu\alpha\beta}\partial_\alpha\partial_\beta\xi$ represents the boundary vortex current and $b = (1/2\pi)\epsilon^{ij}\partial_i\lambda_j$ plays the role of the "charge" canonically conjugate to the phase ξ. As usual, the kinetic terms for the surface modes are added-in *a posteriori* and, correspondingly M is a non-universal mass scale and v is the non-universal propagation speed of the surface modes.

The Hamiltonian of the boundary theory is derived by setting the canonical momenta $\pi_\xi = -(\theta/\pi)b$ and $\pi_{\lambda_2} = -(\theta/4\pi^2)\lambda_1$ (as usual in pure Chern–Simons theories, one has a choice of deciding which of the two components assumes the role of the coordinate and which one is the momentum). Setting the Lagrange multiplier $\lambda_0 = 0$ (Weyl gauge) after noting the Gauss law constraint $b = -\Phi^0$ it implements, gives

$$\mathcal{H}_{\partial M} = \frac{1}{2M}\pi_\xi^2 + \frac{M}{2}v^2(\partial_i\xi)^2, \tag{19.13}$$

which describes a massless degree of freedom, as can be seen from the Hamiltonian equations of motion $\left(\partial_0^2 - v^2\sum_i\partial_i^2\right)\xi = 0$. The gauge field λ_μ is not a dynamical field because there are no "electric fields" appearing in the Lagrangian, and it is expressed entirely in terms of the vortex configuration via the Chern–Simons Gauss law constraint [184]. This constraint, $b = -\Phi^0$, implies that the momentum π_ξ conjugate to the phase ξ is not a pure electric charge, as it would be in superconductors, but the vortex number itself. However, these boundary vortices do carry also unit electric charge $2e$, as can be easily recognized from the electromagnetic coupling $(\theta/2\pi)A_\mu\Phi^\mu$ implied by the θ-term in (19.10). They are thus dyons themselves. Finally, and most importantly, it follows from the formulation (19.11) of the boundary Lagrangian, that these dyons acquire fractional statistics via the boundary Chern–Simons term [79]. In particular, for the relevant case $\theta = 2\pi$ they become *fermions*.

This mechanism can be understood heuristically as follows. As in a superconductor, there is no condensate in a strip of width λ (called for this reason the penetration depth) along the surfaces of the sample. Contrary to a superconductor, however, the "normal state" in this strip does not consist of single electrons, due to Cooper pair splitting. Here there is nothing to split, the charge $2e$ is acquired via the Witten effect. In this strip, thus, we will have just out of condensate charge $2e$ dyons. These interact with the gauge field. However, since we are in a topological state, the gauge field action on this 2D boundary strip cannot be anything other than the Chern–Simons action. Therefore, the gauge field will induce statistical transmutation of these out of condensate boundary dyons.

Thus, the topological limit of strong superinsulators contains only boundary states which are massless fermionic dyons carrying magnetic charge π/e and electric charge $2e$. The canonical structure and the massless character of these boundary fermions rests on the U(1) combined gauge symmetry $\lambda_\mu \to \lambda_\mu + \partial_\mu \chi$ and $\xi \to \xi - \chi$ inherited from the bulk and the time-reversal symmetry due to the angle $\theta = 2\pi$ in the θ-term. Any perturbation that leaves this protecting U(1) $\times$ $\mathbb{Z}_2^T$ symmetry intact does not affect the canonical structure of the boundary fermions. It is to be expected that these symmetry-protected boundary states live on a Chalker–Coddington percolation network [113]. In this internal structure they cannot be localized by impurities. They behave thus as a pure Fermi liquid, giving rise to the characteristic $R_\square \propto T^2$ behaviour. This Fermi-liquid behaviour arises even if the electric charge of the carriers is still $2e$ in perfect accordance with the measurement [171]. An interesting consequence of this mechanism is the possible direct measurement of the dual Ampère law. If we encircle a chunk of material in its pseudogap state by a conducting open ring and we apply a voltage so that an electric current passes through this ring, this current carries with it also magnetic charge. The dual Ampère law predicts then that a voltage should be measured at the two open ends of the ring as a consequence of this magnetic current through it.

The dyonic character of the charge carriers immediately solves also the problem of the nematic behaviour observed in the pseudogap state [175]. The underdoped antiferromagnetic phase of cuprates, holding exactly a single hole per Cu site, realizes a square Heisenberg antiferromagnet. Thus, the electronic symmetry of the lattice is C_4. In such a lattice, the adjacent spins are opposite while the diagonal ones are parallel. As a result, the combined electric/magnetic symmetry of the lattice reduces to C_2. Since the charge carriers are dyons, carrying both electric and magnetic charge, they feel the combined C_2 symmetry and not only the electronic C_4 symmetry. Hence nematicity emerges, with the nematic director going along the CuO_2 lattice diagonals. This essential characteristic of the PG state [175, 176] goes hand-in-hand with the T^2 resistance.

Let us finally turn to the mystery of the strange metal in the overdoped region [178, 179]. If a magnetic field is applied in this region which is strong enough to destroy superconductivity, a linear resistance $R_\square \propto T$ is observed down to the lowest temperatures. A linear resistance typically characterizes metals above the Debye temperature T_D, where it is caused by scattering with phonons (see, e.g. [185]). At the temperatures $T \ll T_D$ where the strange metal is observed, however, phonons are frozen and a metal would have a resistance scaling like $R_\square \propto T^5$ [185]. Note also that, contrary to assertions sometimes found in the literature, the slope of the linear resistance at very low temperatures is clearly different from the slope in the normal metal above T_c [179]. The strange metal at very low temperatures is something entirely different than the normal metal above T_c.

The solution of the strange metal puzzle lies again in the dyon condensate. While a strong magnetic field does indeed destroy the pure charge condensate, it does not destroy the coexisting (charged) magnetic monopole condensate. First of all, the out-of-condensate Cooper pairs of charge $2e$ feel the surviving magnetic monopoles of magnetic charge π/e and, as we derived in Chapter 16, are turned by them into fermions. Secondly, the metastable bubbles of this surviving condensate are characterized by low-energy coherent fluctuations of the condensate, called Bogoliubov excitations, or bogolons. In hybrid fermion–boson systems like the present one, the interaction of fermionic charges with these bogolons is known to result in a linear $R_\square \propto T$ resistance behaviour [186] for temperatures above another characteristic temperature, the Bloch–Grüneisen temperature $T_{\mathrm{BG}} = 2sk_{\mathrm{F}}/k_{\mathrm{B}}$ [185], where k_{F} is the Fermi momentum of the fermions and s the sound velocity of the bogolons. Bogolons simply play the same role for $T > T_{\mathrm{BG}}$ as phonons do for $T > T_{\mathrm{D}}$. In high-T_c cuprates the zero-temperature superfluid density n_{s} and the effective "Fermi energy" $\epsilon_{\mathrm{F}}^* \simeq n_{\mathrm{s}}\xi/4m^*$ associated with fermionic Cooper pairs are relatively small [187] (here ξ is the superconducting coherence length, and m^* is the effective mass of the carriers). Accordingly, the Bloch–Grüneisen temperature T_{BG} is very low, especially if the bogolon sound velocity s is also small. We expect thus a linear-in-T resistance behaviour down to the lowest temperatures. The strange metal is made of statistically transmuted Cooper pairs interacting with long-wave excitations of the monopole condensate.

Note, finally that, at high magnetic fields B, the number of available degenerate ground states in an area A is $N = (eB/2\pi)$. This is the number of available scattering centres at $T = 0$ and the same reasoning that leads to the linear-in-T resistance leads also to the observed [179] linear-in-B behaviour of the intercept at $T = 0$. In other words, at high fields the role of the temperature T is taken by the cyclotron frequency $\omega_{\mathrm{c}} = eB/m^*$.

To conclude this chapter we derive the *oblique Meissner effect* of the dyon condensate in the pseudogap state. The electromagnetic response of the oblique superinsulator (19.8) implies the following modification to the equations of motion (17.14),

$$\partial_\mu H^{\mu\alpha\beta} + \frac{\Lambda^2}{f^2}\left(B^{\alpha\beta} - \frac{f^2}{2\pi}\frac{\theta}{2\pi}\tilde{B}^{\alpha\beta}\right) = 0. \tag{19.14}$$

These new equations, in turn, are equivalent to the following modified London equations for the oblique superinsulator (for stationary field configurations)

$$\frac{2\pi}{f}\,\partial_0 \mathbf{j}_g = \frac{\Lambda^2}{f^2}\left(\mathbf{B} + \frac{f^2}{2\pi}\frac{\theta}{2\pi}\mathbf{E}\right),$$
$$\frac{2\pi}{f}\nabla \wedge \mathbf{j}_g = \frac{\Lambda^2}{f^2}\left(\mathbf{E} - \frac{f^2}{2\pi}\frac{\theta}{2\pi}\mathbf{B}\right), \tag{19.15}$$

where $\mathbf{j}_g$ is the monopole number current (17.22). Let us now combine the second of these equations with the (static) Ampère equations

$$\nabla \wedge \mathbf{B} = f\frac{\theta}{2\pi}\mathbf{j}_g,$$
$$\nabla \wedge \mathbf{E} = -\frac{2\pi}{f}\mathbf{j}_g, \tag{19.16}$$

where we have used the fact that the electric number current is $\mathbf{j} = (\theta/2\pi)\mathbf{j}_g$ since it is carried by charged monopoles. Taking the curl of the second equation in (19.15) and using (19.16) we get (for sourceless currents)

$$\left(\nabla^2 - \frac{1}{\lambda_\theta^2}\right)\mathbf{j}_g = 0, \tag{19.17}$$

where $\lambda_\theta = 1/vm_\theta$ with m_θ given in (19.9). This shows that currents are screened on the scale of the penetration depth λ_θ in the interior of the oblique superinsulator. The residual surface currents screen external electromagnetic fields, as in a traditional superconductor. However, as is evident from (19.15), applied fields for which

$$\mathbf{E} = \frac{f^2}{2\pi}\frac{\theta}{2\pi}\mathbf{B}, \tag{19.18}$$

are not screened and can penetrate the oblique superinsulator. Let us now decompose any combination of electromagnetic fields in the orthogonal basis

$$\begin{pmatrix} \frac{f^2}{2\pi}\frac{\theta}{2\pi} \\ 1 \end{pmatrix}, \quad \begin{pmatrix} -1 \\ \frac{f^2}{2\pi}\frac{\theta}{2\pi} \end{pmatrix}, \tag{19.19}$$

where the upper component is the electric one and the lower the magnetic one. Components along the first basis vector can penetrate the superinsulator, components along the second basis vector are expelled. Let us apply, for example, an external magnetic field $\mathbf{B}$. We can decompose the corresponding vector $(0, \mathbf{B})$ into its components along the above orthogonal basis and the second component will be expelled. Therefore the oblique superinsulator will react by generating compensating magnetic and electric fields

$$\mathbf{B}_{\mathrm{com}} = -\frac{\left(\frac{f^2}{2\pi}\frac{\theta}{2\pi}\right)^2}{1 + \left(\frac{f^2}{2\pi}\frac{\theta}{2\pi}\right)^2}\mathbf{B},$$

$$\mathbf{E}_{\mathrm{com}} = \frac{\frac{f^2}{2\pi}\frac{\theta}{2\pi}}{1 + \left(\frac{f^2}{2\pi}\frac{\theta}{2\pi}\right)^2}\mathbf{B}.$$

(19.20)

In the exact topological limit $f^2 \to \infty$ external magnetic fields are completely screened and no electric field is generated. In the physical regime $f^2/2\pi \gg 1$, however, the oblique superinsulator reacts to an applied magnetic field by producing a small electric field $\mathbf{E} \propto \frac{2\pi}{\theta}(2\pi/f^2)\mathbf{B}$. Equations (19.20) describe the oblique Meissner effect.

Chapter 20

Real-space electron pairing by magnetic monopoles

Electron pairing in high-T_c superconductors (for a review see [42, 172]) is due to a different mechanism than the standard s-wave Bardeen–Cooper–Schrieffer (BCS) phonon-mediated attraction (for a review see [5]). Typical existing proposals for a different mechanism, however, still focus on BCS-type pairing, either of different objects or mediated by different massless fields. In Anderson's famed resonating valence bond (RVB) model [188] and variants thereof [189] the BCS wave function is a superposition of neutral couples of spin-singlet electrons that become charged and superconduct upon doping (for a review see [190]). Otherwise, BCS-type pairing by spin density waves [191] and by bosonic collective excitations of the fermions that become massless at a quantum critical point have been considered [192,193]. Whatever the correct mechanism, however, it must guarantee that electron pairs in high-T_c materials are localized on scales of a few nm by very strong interactions, a situation very different from the weak BCS pairing in standard superconductors [194], as illustrated in Fig. 20.1. The phenomenology of high-T_c materials points to the formation of localized electron pairs that subsequently Bose condense at a lower temperature [195].

The effective field theory of high-T_c superconductivity presented in Chapter 19 is predicated on the type of emergent, self-organized granularity typical of 2D superconducting films [35] and recently observed also in a 3D superconducting material [37], in which real-space bubbles of superconducting condensate form and become phase coherent by tunnelling when sufficiently near each other. To support this picture of high-T_c superconductivity we need, of course, a pairing mechanism which is totally different from the usual momentum-space, boson exchange. It should be a real-space mechanism of pairing in which granules of condensate form around random

195

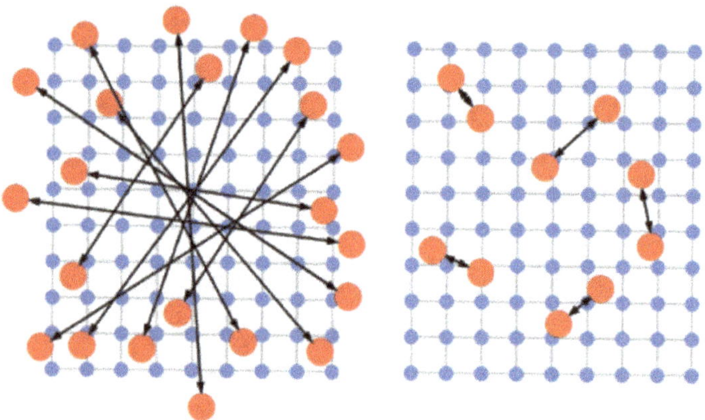

Fig. 20.1. (Left) Delocalized, weak-coupling BCS electron pairs in conventional low-T_c superconductors. (Right) Localized, strong-coupling electron pairs in high-T_c materials. From A. Bussmann-Holder and H. Keller, High-temperature superconductors: Underlying physics and applications, *Z. Naturforsch.* (2019), with permission by the authors and the publisher. ©De Gruyter (2019).

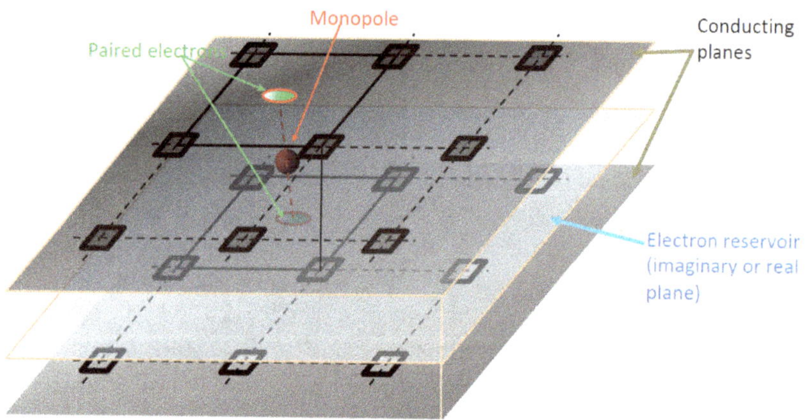

Fig. 20.2. Real-space pairing by a magnetic monopole between the two conducting planes making up a unit layer of a cuprate superconductor. The squares illustrate the arrangement of atoms on these planes.

seeds, in full accord with the idea of preformed, very localized pairs that Bose condense at a lower temperature. As we now show, such a mechanism is provided again by magnetic monopoles, forming between the two conducting planes of a unit layer of cuprate superconductors [43], as illustrated in Fig. 20.2.

Of course, the question arises about what is the origin of these monopoles. One possibility, discussed in detail in Chapter 21, is that they appear as real magnetic monopoles in loop current Mott insulators. Another is that they are effective, so-called pseudo magnetic monopoles. It has been known for a long time that defects in condensed matter systems can be described by effective gauge fields [57]. In graphene sheets, strains, dislocations, and curvature (ripples) are equivalent to an effective gauge field coupled to the low-lying electronic degrees of freedom [196,197] which typically amounts to a magnetic field perpendicular to the sheet (for a review see [198]). This is typically called pseudo magnetic field since it does not correspond to a real applied magnetic field but is, rather, an effective description of the geometric curvature effects induced by the defects. Most interestingly, it has been recently shown that the curvature of graphene nanobubbles is equivalent to a pseudo magnetic monopole at the centre of the bubble [199]. Finally, also in cuprates, hedgehog defects in the magnetic order can appear as effective monopoles to itinerant electrons due to Berry phases. Here we shall not dwell further on the exact origin of these monopoles but, rather, we will focus on the pairing mechanism they induce.

Consider the quantum mechanical problem of two electrons interacting by a spherically symmetric repulsive $(1/r)$ Coulomb potential and a short-range repulsion $V_R(r)$. The decomposition of the wave function into spherical harmonics provides an additional $(\ell(\ell+1)/r^2)$ repulsive centrifugal barrier for states with angular momentum $\ell > 0$. If the short-range repulsion $V_R(r)$ is stronger than the centrifugal barrier, it pays for the system to settle in a sufficiently high angular momentum state. Due to rotational symmetry, the z-component m, $|m| \leq \ell$, of the angular momentum falls out of the Hamiltonian. What happens, however, if a spherical-symmetry-breaking mechanism encoded in a *vector potential* makes the Hamiltonian explicitly dependent on the z-component of angular momentum? On dimensional grounds this would cause a further $(1/r^2)$ term in the effective potential which modifies the centrifugal barrier. Such a term, however, could in principle have either sign. In case of a negative sign, it can even counterbalance the centrifugal barrier so that an attractive $(1/r^2)$ potential arises at intermediate distances, before the Coulomb repulsion takes over. In this case, a potential well could form, giving rise to a discrete spectrum of bound states with finite angular momentum.

Does such a mechanism exist? The answer is yes, this mechanism is provided by *magnetic monopoles* [17]. As we have shown in Chapter 16, in the presence of a magnetic monopole of strength g, an electron acquires an additional angular momentum $L_\mathrm{M} = -(eg/4\pi)\hat{\mathbf{r}}$ where $\hat{\mathbf{r}}$ is the unit vector pointing from the monopole location to the electron one. The Dirac quantization condition $eg = 2\pi n$, $n \in \mathbb{Z}$, is the requirement that this additional angular momentum contribution, originating from the interplay of the electric and magnetic fields of the two point particles, matches the spectrum imposed by the rotation group. This is exactly a spherical-symmetry-breaking contribution to angular momentum, since it singles out the vector joining the monopole to the electron. As we now show, this monopole-induced angular momentum contribution can indeed counterbalance the centrifugal barrier, giving rise to bound pairs that can Bose condense. The optimal angular momentum of the pair depends on the monopole density. Lower densities favour higher angular momenta and *vice versa*.

In Chapter 16 we have also shown that monopoles change the statistics of the electrons: these become bosons themselves for $eg/2\pi$ odd and remain fermions for $eg/2\pi$ even (for a review see [17]). The centrifugal barrier cancellation can take place only for odd values of $eg/2\pi$. Magnetic monopoles, thus, induce pairing of electrons that can Bose condense into localized droplets around the monopole location. Because of the Dirac quantization condition, magnetic monopoles are very heavy excitations, with a mass $m_\mathrm{M} \propto 1/\alpha$, with α the fine structure constant. Therefore, these droplets are essentially fixed in space and constitute the random seeds of the granular Bose condensate.

High-T_c materials are Janus-like as far as their dimensionality is concerned. On one side, charge transport is two-dimensional (2D) [200, 201], on the other side, topological aspects, like the magnetoelectric effect in the pseudogap state [174–177], require the full three dimensions (3D). This is the reason why even a monolayer of these materials, consisting of *two conducting planes*, retains its high transition temperature [173], while really 2D films of similar materials have a much lower transition temperature. We shall consider the situation in which an infinitely heavy fundamental magnetic monopole, the 3D effect, sits in between the two conducting planes, see Fig. 20.2, while electrons are allowed to move only along these horizontal planes, corresponding to 2D charge transport.

Let us thus consider the quantum mechanical problem of two electrons of charge e with short-range repulsion in presence of a Dirac magnetic monopole of magnetic charge g. Of course, this is a three-body problem, which is not easy to solve analytically in full generality. To proceed we shall thus consider the easier problem of an infinitely heavy magnetic monopole sitting fixed in the centre of mass of the two-electron system. This reduces the model to the one-body problem of an electron of reduced mass $m/2$ in the external field of the fixed magnetic monopole, which is amenable to an analytical, albeit approximate solution.

The derivation of the potential well goes along the same line as the derivation of the statistical transmutation in Chapter 16. As we have stressed there, Dirac monopoles are singular configurations, with a Dirac string attached to them. If the string lies along the negative z-axis, the gauge potential $\mathbf{A}_u$ of the monopole is [17]

$$\mathbf{A}_u = f_u(r,\theta)\,\hat{\varphi},$$
$$f_u(r,\theta) = \frac{g}{4\pi r}\frac{1-\cos(\theta)}{\sin(\theta)}, \tag{20.1}$$

where r, θ, and φ denote the usual spherical coordinates and $\hat{\varphi}$ is the unit vector in φ direction. The singularity at $\theta = \pi$ is evident from this expression. To solve problems in the field of a Dirac monopole one cannot use, thus a single set of coordinates for the whole sphere but one must use the Wu–Yang formalism [152] and cover the sphere with a so-called atlas, supplemented by gauge transformation conditions between the different maps. The simplest atlas consist of two maps, the northern hemisphere $0 \le \theta \le \pi/2 + \epsilon$, with the gauge potential (20.1) and the lower hemishpere $\pi/2 - \epsilon \le \theta \le \pi$, with the gauge transformed gauge potential

$$\mathbf{A}_l = f_l(r,\theta)\,\hat{\varphi},$$
$$f_l(r,\theta) = -\frac{g}{4\pi r}\frac{1+\cos(\theta)}{\sin(\theta)}, \tag{20.2}$$

corresponding to the same magnetic monopole at the origin but with the Dirac string now along the positive z-axis,

$$\mathbf{A}_l = \mathbf{A}_u - \nabla\left(\frac{g}{2\pi}\varphi\right). \tag{20.3}$$

In both hemispheres the gauge potential is now regular and the corresponding Pauli equation can be solved. The price to pay are gauge transformation conditions for the wave function on the overlap $[\pi/2 - \epsilon, \pi/2 + \epsilon]$ of the two hemispheres.

The Hamiltonian for two electrons of charge e and mass m in the field of a fixed magnetic monopole of magnetic charge g at the origin is given by

$$H = \frac{1}{2m}\left(\mathbf{p}_1 - e\mathbf{A}(\mathbf{x}_1)\right)^2 + \frac{1}{2m}\left(\mathbf{p}_2 - e\mathbf{A}(\mathbf{x}_2)\right)^2$$

$$- \frac{e}{m}\mathbf{s}_1 \cdot \mathbf{B}(\mathbf{x}_1) - \frac{e}{m}\mathbf{s}_2 \cdot \mathbf{B}(\mathbf{x}_2)$$

$$+ V_R\left(|\mathbf{x}_1 - \mathbf{x}_2|\right) + V_C\left(|\mathbf{x}_1 - \mathbf{x}_2|\right), \tag{20.4}$$

where $\mathbf{A}$ is the monopole gauge potential (20.1) or (20.2), $\mathbf{s}_{1,2}$ denote the spin vectors of the two electrons, $V_C(r) = e^2/4\pi\varepsilon r$ is the repulsive Coulomb potential with ε the relative dielectric permittivity of the material, and $V_R(r)$ denotes the short-range repulsion. To establish the existence of possible bound states it is sufficient to analyze the relative problem. We introduce thus centre-of-mass and relative coordinates $\mathbf{R} = (\mathbf{x}_1 + \mathbf{x}_2)/2$ and $\mathbf{r} = (\mathbf{x}_1 - \mathbf{x}_2)$ and we set $\mathbf{R} = 0$. To make the model amenable to an analytical solution we shall moreover make the simplifying assumption that the infinitely heavy, external magnetic monopole sits exactly in the centre of mass of the two-electron system. The time-independent Pauli equation becomes then

$$\left[\frac{-1}{2m}\left(\nabla - ie\mathbf{A}\left(\frac{\mathbf{r}}{2}\right)\right)^2 + \frac{-1}{2m}\left(\nabla + ie\mathbf{A}\left(\frac{-\mathbf{r}}{2}\right)\right)^2 - \frac{|eg|}{2\pi m|\mathbf{r}|^2}\right.$$

$$\left. + V_R(|\mathbf{r}|) + V_C(|\mathbf{r}|)\right]\psi = E\psi, \tag{20.5}$$

where we have specialized to a total spin 0 state in which the spin of each electron has a hedgehog configuration parallel or antiparallel to the monopole magnetic field, depending on the sign of g. Using the fact that the monopole gauge potentials (20.1) and (20.2) are divergenceless, we can simplify this to

$$\left[\frac{-1}{m}\nabla^2 + \frac{ie}{2m}\left(\mathbf{A}\left(\frac{\mathbf{r}}{2}\right) \cdot \nabla - \mathbf{A}\left(\frac{-\mathbf{r}}{2}\right) \cdot \nabla\right)\right.$$

$$+ \frac{e^2}{2m}\left(\mathbf{A}^2\left(\frac{\mathbf{r}}{2}\right) + \mathbf{A}^2\left(\frac{-\mathbf{r}}{2}\right)\right) - \frac{|eg|}{2\pi m|\mathbf{r}|^2}$$

$$\left. + V_R(|\mathbf{r}|) + V_C(|\mathbf{r}|)\right]\psi = E\psi. \tag{20.6}$$

Of course this is still a formal expression. To be precise we have to formulate two Pauli equations, one for the upper hemisphere and one for the lower hemisphere, as explained above. Let us begin with the upper hemisphere, denoted by the subscript "u". Since we restrict to values $0 \leq \theta \leq \pi/2 + \epsilon$, the arguments of the second gauge potentials $\mathbf{A}$ in (20.6) inevitably refer to the lower hemisphere, denoted by subscripts "l". Therefore, we have to use (20.1) for the first instance of the gauge potential and (20.2) for the second one. This gives

$$\left[\frac{-1}{m}\nabla^2 + \frac{ie}{2m}\left(f_u\left(\frac{r}{2},\theta\right) - f_l\left(\frac{r}{2},\pi-\theta\right)\right)\frac{1}{r\sin(\theta)}\frac{\partial}{\partial\varphi}\right.$$
$$+ \frac{e^2}{2m}\left(f_u^2\left(\frac{r}{2},\theta\right) + f_l^2\left(\frac{r}{2},\pi-\theta\right)\right) - \frac{|eg|}{2\pi mr^2}$$
$$\left. + V_R(r) + V_C(r)\right]\psi_u = E\psi_u. \tag{20.7}$$

Repeating the same reasoning for the lower hemisphere $\pi/2 - \epsilon \leq \theta \leq \pi$, we obtain the second Pauli equation

$$\left[\frac{-1}{m}\nabla^2 + \frac{ie}{2m}\left(f_l\left(\frac{r}{2},\theta\right) - f_u\left(\frac{r}{2},\pi-\theta\right)\right)\frac{1}{r\sin(\theta)}\frac{\partial}{\partial\varphi}\right.$$
$$+ \frac{e^2}{2m}\left(f_l^2\left(\frac{r}{2},\theta\right) + f_u^2\left(\frac{r}{2},\pi-\theta\right)\right) - \frac{|eg|}{2\pi mr^2}$$
$$\left. + V_R(r) + V_C(r)\right]\psi_l = E\psi_l. \tag{20.8}$$

Using (20.1) and (20.2), we obtain finally the explicit expressions of our pair of Pauli equations,

$$\left[\frac{-1}{m}\nabla^2 + i\frac{(eg/2\pi)}{mr^2}\frac{1-\cos(\theta)}{\sin^2(\theta)}\frac{\partial}{\partial\varphi} + \frac{(eg/2\pi)^2}{mr^2}\left(\frac{1-\cos(\theta)}{\sin(\theta)}\right)^2\right.$$
$$\left. - \frac{|eg|}{2\pi mr^2} + V_R(r) + V_C(r)\right]\psi_u = E\psi_u,$$

$$\left[\frac{-1}{m}\nabla^2 - i\frac{(eg/2\pi)}{mr^2}\frac{1+\cos(\theta)}{\sin^2(\theta)}\frac{\partial}{\partial\varphi} + \frac{(eg/2\pi)^2}{mr^2}\left(\frac{1+\cos(\theta)}{\sin(\theta)}\right)^2\right.$$
$$\left. - \frac{|eg|}{2\pi mr^2} + V_R(r) + V_C(r)\right]\psi_l = E\psi_l. \tag{20.9}$$

The presence of the magnetic monopole is reflected in three new terms, the first two of which, as anticipated, break the spherical symmetry of the original Coulomb problem. The first embodies the monopole-induced additional contribution to the z-axis component of the angular momentum: it has a different sign in the upper and lower hemispheres since, as we discussed above, it points from the monopole to the electrons and the monopole sits exactly in the middle. Its coefficient is only weakly dependent on θ since it varies from 1 on the equator to 1/2 at the poles. The second new term, instead, is a repulsive term concentrated around the equator and vanishing near the poles. Finally, the third new term is the magnetic attraction due to the electron magnetic moments.

Since the vector gauge potentials in the lower and upper hemispheres are gauge transforms of each other, we must impose the Wu–Yang gauge conditions [152] also on the wave functions in the overlap region $[\pi/2 - \epsilon, \pi/2 + \epsilon]$ of the maps of the atlas,

$$\psi_l = e^{-i\frac{eg}{2\pi}\varphi}\psi_u. \tag{20.10}$$

We can thus make the ansatz

$$\psi_u\left(r,\theta,\varphi\right) = e^{+i\frac{eg}{4\pi}\varphi}F_u(r,\theta,\varphi),$$
$$\psi_l\left(r,\theta,\varphi\right) = e^{-i\frac{eg}{4\pi}\varphi}F_l(r,\theta,\varphi). \tag{20.11}$$

Because an exchange of the two electrons involves necessarily also a swap of hemispheres, the exchange operator on the wave function must take into account the Wu–Yang gauge transformation. Therefore, for $eg/2\pi$ an odd integer the exchange implies a factor (-1) and the statistics of the electrons is changed to bosons. For $eg/2\pi$ an even integer, instead there is no additional (-1) factor and the statistics of the individual electrons remains fermionic. Correspondingly, for $eg/2\pi$ an odd integer the additional gauge factor is a double covering representation of 2π rotations, while it is single-valued for $eg/2\pi$ an even integer. This is the statistical transmutation induced by magnetic monopoles (for a review see [17]). Independently of this statistical transmutation of the individual components, however, the total spin 0 pair is a boson.

To make contact with the physical setting of high-T_c materials, as explained above, we now constrain the electrons to move on parallel horizontal planes at $z = \pm\delta$ with the monopole at the origin. In this case we can dispense with monopole harmonics [152] and the Laplace operator becomes

$$\nabla^2 = \frac{\partial^2}{\partial x^2} + \frac{1}{x}\frac{\partial}{\partial x} + \frac{1}{x^2}\frac{\partial^2}{\partial \varphi^2}, \tag{20.12}$$

where x denotes the radial distance on the two planes. Since we will be interested only in small values of x and the second new term in (20.9), $O(\theta^2)$, is subdominant with respect to the first one, $O(1)$, near the poles, we will henceforth neglect it. We thus specialize our ansatz to

$$
\begin{aligned}
F_u(r,\theta,\varphi) &= \mathrm{e}^{-i\ell\varphi} F(x), \\
F_l(r,\theta,\varphi) &= \mathrm{e}^{+i\ell\varphi} F(x),
\end{aligned}
\tag{20.13}
$$

where r and θ are bound by the condition $x = r\sin\theta$ and ℓ is the total angular momentum, $2\ell \in \mathbb{N}$. Note that the z-components of the angular momentum of the electrons on the two planes cancel out, but the total angular momentum ℓ can well be different from zero. It is this value that indicates how much the axis of the composite wave function is tilted with respect to the z-axis.

Because of the denominator $r^2 = \delta^2 + x^2$, the monopole effects reduce to a constant at very short distances $x \ll \delta$. Suppose we neglect entirely the monopole and use the familiar cylindrical harmonics decomposition of the Laplacian for these short distances. We then obtain a single radial equation for both planes,

$$
\left[\frac{-1}{m}\left(\frac{1}{x}\frac{\partial}{\partial x}\left(x\frac{\partial}{\partial x} \right) \right) + V_R(x) + \frac{\ell^2}{mx^2} - \frac{(eg/2\pi)^2}{4m(\delta^2 + x^2)} - \frac{|eg/2\pi|}{m(\delta^2 + x^2)} \right.
$$
$$
\left. + V_C(x) \right] F(x) = EF(x),
\tag{20.14}
$$

where $V_C(x) = \alpha/\varepsilon\sqrt{\delta^2 + x^2}$, with $\alpha = e^2/4\pi \approx 1/137$ the fine structure constant. The short-range repulsion models the quantum statistical pressure of electrons when they are squeezed by the two conducting planes. Its exact form does not matter, however, it forces electrons to fall into non-zero angular momentum states $\ell > 0$ in order to avoid the energy price to be extremely close: $F(x) \propto x^\ell$ for $x \ll \delta$ and the higher ℓ, the more the wave function is suppressed at the origin. However, for $eg/2\pi = \pm 2\ell$ the resulting repulsive centrifugal barrier gets cancelled at distances $x > O(\delta)$ by the monopole-induced angular momentum, and only an attractive interaction due to the electron magnetic moment survives. Of course, in the exact solution the complete cancellation is evident simply by combining (20.11) and (20.13). A potential well forms between the two relevant scales δ and $a = \varepsilon/m\alpha$, where the Coulomb repulsion takes over, and the electrons form pairs that can Bose condense in a droplet localized around the positions of the heavy monopoles. What is the optimal value of ℓ? On one side, increasing ℓ makes the magnetic moment attraction stronger; on the other side, however, the higher the monopole charge, the heavier they are and the higher the energy cost of creating them between the planes. Given the high mass of the monopoles, it is to be expected that at small monopole densities larger values of ℓ are favoured and *vice versa*.

The formation of the potential well is again an interplay of dimensionality. The angular momentum of the electrons constrained to move on the two planes is a 2D effect. The additional angular momentum due to the monopole and the magnetic moment interactions, however, are 3D effects, since they point from the monopole at the centre to the locations of the electrons on the planes. For sufficiently high distances the additional angular momentum cancels out the centrifugal barrier and the magnetic moment interaction causes the overall attraction.

The potential well is determined by the combination of the magnetic moment attraction with the short-range repulsion. This is supposed to model the quantum statistical pressure of electrons of inter-layer atoms when they are squeezed by the two conducting planes. As such it should be scale-free, i.e. a $1/x$ potential. Therefore, in general, both the position of the minimum of the potential well and the bound state energy are functions of the two scales δ and a. If these are comparable, however, there remains only one single scale determining all properties of a single monopole well, which we take as δ. By multiplying (20.14) overall by m we obtain a right-hand side that depends only on mE and a left-hand side that depends only on the scale δ. Therefore, the the binding energy E_0 is

$$E_0 = O\left(\frac{\hbar^2}{m\delta^2}\right) = O\left(E_{\rm F}\right), \tag{20.15}$$

where we have reinstated physical units and we have used the typical inter-plane distance $\delta \approx k_{\rm F}$, with $k_{\rm F}$ the Fermi wavevector, and $E_{\rm F}$ the Fermi energy.

This is the whole story for a single monopole. The binding energy, however, is typically increased in presence of a finite density of monopoles. To see how this comes about for generic potential wells let us start from the generic Schrödinger equation for a particle of mass m in an external potential well $U(\mathbf{r})$

$$\left(-\frac{1}{2m}\Delta + U(\mathbf{r})\right)\psi(\mathbf{r}) = E\psi(\mathbf{r}). \tag{20.16}$$

For simplicity of presentation we will consider a square well such that

$$U(\mathbf{r}) = \begin{cases} -U, & |\mathbf{r}| < R, \\ 0, & |\mathbf{r}| > R, \end{cases}$$
$$\lim_{R \to 0} UR^2 = \text{const.}, \tag{20.17}$$

and we shall assume that there exists a bound state with energy $E = -E_0$, such that $E_0/U \to 0$ for $R \to 0$. In the limit $R \to 0$ this reduces to a δ-potential. We will denote by ψ_1 the solution of the Schrödinger equation for $|\mathbf{r}| < R$ and by ψ_2 the corresponding solution outside the potential well, $|\mathbf{r}| > R$. At $r = R$ we must then impose the continuity conditions

$$\psi_1 = \psi_2, \quad \frac{d\psi_1}{dr} = \frac{d\psi_2}{dr}. \tag{20.18}$$

We now introduce the usual decomposition $\psi(\mathbf{r}) = R_\ell(r)Y_{\ell,m}(\theta, \varphi)$ into a product of a radial function and spherical harmonics $Y_{\ell,m}(\theta, \varphi)$. The equation for the radial function reduces to

$$\frac{1}{r^2}\frac{d}{dr}\left(r^2\frac{dR_\ell}{dr}\right) + \left[2m(E - U) - \frac{\ell(\ell+1)}{r^2}\right]R_\ell = 0. \tag{20.19}$$

Introducing the new function $\chi_\ell(r) = rR_\ell(r)$ this translates to

$$\chi_\ell'' - \left[\kappa^2 + \left(V(r) + \frac{\ell(\ell+1)}{r^2}\right)\right]\chi_\ell = 0, \tag{20.20}$$

where $V(r) = 2mU(r)$ and $\kappa = \sqrt{2mE_0}$ and we have denoted radial derivatives by an apostrophe. Let us now specialize, for simplicity, to the s-wave case $\ell = 0$. Outside the potential well the equation reduces to

$$\chi_2'' - \kappa^2\chi_2 = 0, \tag{20.21}$$

with solution

$$\chi_2(r) = \mathrm{e}^{-\kappa r}. \tag{20.22}$$

Inside the potential well, instead, we have

$$\chi_1'' + \left(V - \kappa^2\right)\chi_1 = 0, \tag{20.23}$$

with $V = 2mU > \kappa^2$ and with solution

$$\chi_1(r) = A\sin(kr + \delta(k)), \tag{20.24}$$

and $k = \sqrt{V - \kappa^2}$.

The matching conditions can now be presented as

$$\chi_1(R) = \chi_2(R),$$
$$\frac{d\ln\chi_1}{dr}\bigg|_R = \frac{d\ln\chi_2}{dr}\bigg|_R = -\kappa. \qquad (20.25)$$

In the limit $R \to 0$ of a δ-potential, these matching equations reduce to

$$1 = A\sin(\delta(k)),$$
$$k\cot(\delta(k)) = -\kappa. \qquad (20.26)$$

Together with the normalization condition $\int d^3\mathbf{r}\,|\psi(\mathbf{r})|^2 = 1$, these equations determine the two parameters A and δ and the bound state energy κ, the crucial point being that this bound state energy is determined by the quantity

$$z = \frac{d\ln\chi}{dr}\bigg|_0 = -\kappa, \qquad (20.27)$$

where we have left out the subscript "2" on the wave function since we have taken the limit $R \to 0$. A potential well can bind a particle (in the s-wave) only if its z value is negative.

On this basis, let us now derive how the bound state energy of the particle is changed to $E_2 = -\kappa_2^2/2m$ when there are two potential wells with given, fixed z located at $\mathbf{r}_1$ and $\mathbf{r}_2$ (from now on the numerical subscripts refer to the quantity of potential wells). For not-too-close wells, so that interference effects between the two can be neglected, we expect the solution to take the form

$$\psi = \frac{e^{-\kappa_2|\mathbf{r}-\mathbf{r_1}|}}{|\mathbf{r}-\mathbf{r_1}|} + \frac{e^{-\kappa_2|\mathbf{r}-\mathbf{r_2}|}}{|\mathbf{r}-\mathbf{r_2}|}. \qquad (20.28)$$

Near $\mathbf{r} = \mathbf{r}_1$ we expand the first exponent,

$$\psi = \frac{1}{|\mathbf{r}-\mathbf{r_1}|} - \kappa_2 + \frac{e^{-\kappa_2 r_{12}}}{r_{12}}, \qquad (20.29)$$

where $r_{12} = |\mathbf{r}_1 - \mathbf{r}_2|$. We can then compute the z value of each individual potential,

$$z = \frac{e^{-\kappa_2 r_{12}}}{r_{12}} - \kappa_2. \qquad (20.30)$$

This demonstrates that, even if one potential well alone is too weak to bind a particle, i.e. $z > 0$, the presence of a second well close enough is sufficient to create an overall bound state for the system: the condition is $r_{12} < 1/z$.

Consider now N potential wells randomly, but evenly distributed, with the density ρ and let

$$\psi = \sum_i \frac{e^{-\kappa_N |\mathbf{r} - \mathbf{r}_i|}}{|\mathbf{r} - \mathbf{r}_i|}. \tag{20.31}$$

Then, the equation for κ_N becomes

$$z = \sum_{j \neq i} \frac{e^{-\kappa_N |\mathbf{r}_i - \mathbf{r}_j|}}{|\mathbf{r}_i - \mathbf{r}_j|} - \kappa_N, \tag{20.32}$$

and, because the distribution of the potential wells is uniform, this is independent of i. Replacing the sum by an integral we obtain

$$z = -\kappa_\rho + \rho \int \frac{e^{-\kappa_\rho r}}{r} d^3 \mathbf{r} = -\kappa_\rho + \frac{4\pi\rho}{\kappa_\rho^2}. \tag{20.33}$$

There are two length scales in the problem, the Compton wavelength of the bound state, $1/\kappa_\rho$, and the typical distance between potential wells, $\rho^{-1/3}$. Correspondingly, we have two extreme regimes. In the very low density regime $1/\kappa_\rho \ll \rho^{-1/3}$, the second term on the right-hand side of (20.33) can be neglected and the single-well energy dominates (of course if a single well is strong enough to bind a particle, i.e. $z < 0$). In the very high density regime $1/\kappa_\rho \gg \rho^{-1/3}$, the first term on the right-hand side of (20.33) can be neglected, which means that $z > 0$ and a single well cannot bind the particle anymore. In this regime the bound state energy is dominated by the potential well density ρ,

$$E_\rho = -\frac{4\pi\rho}{2mz}. \tag{20.34}$$

In 2D the binding energy for two potential wells was computed analytically in [132]. The localization length $\ell_\parallel$ within the planes is $\ell_\parallel \approx \delta$. When the monopole density increases, the binding energy E_0 at first increases [202]. When the monopole density reaches a level such that the inter-monopole distance $d = (\delta\rho)^{-1/2} > \delta$, however, the "flat" potential well bottoms start to overlap and, due to 2D Lifshitz localization [202], it is the fluctuations in the monopole density rather than the single potential wells that become determinant for localizing pairs. In this regime the binding energy E_0 starts again to decrease [203]. The optimal inter-monopole distance for maximum binding energy will thus be reached at $d \approx \delta$. This prediction for the optimal monopole density is in accord with recent experimental data of [204] showing that the highest transition temperature in carbonaceous sulfur is indeed achieved at some optimal distance between the conducting planes.

Magnetic monopoles are thus the nucleation seeds for superconducting droplets of typical size $\xi(T = 0) = \delta$. When the temperature is raised, of course, the droplet size $\xi(T)$ can increase. Global superconductivity sets in at the temperature at which sufficient monopoles have formed so that the superconducting droplets are near enough for pairs to tunnel from one to the other. The critical temperature is given by the Ioffe–Larkin formula [205] for T_c in highly inhomogeneous superconductors,

$$k_B T_c = \omega \; e^{-0.89 \frac{d}{\xi(T_c)}}, \tag{20.35}$$

where ω is the attempt rate in the matrix element $t = \omega \, \exp(-d/\xi)$ for tunnelling of a pair from one droplet to the next. To estimate the maximum critical temperature we take the optimal monopole density that provides the strongest binding, i.e. $d = \delta$. To make this estimate as conservative as possible, however, we choose the smallest possible droplet size, i.e. $\xi = \delta$. The attempt rate for tunneling between droplets of size s over a distance d is [203]

$$\omega = \sqrt{\frac{8\delta}{\pi d}} \left(\frac{\hbar^2}{m\delta^2} \right). \tag{20.36}$$

Taking $d = \delta$ gives the estimate

$$T_c \approx 0.65 \frac{\hbar^2}{2m\delta^2 k_B}. \tag{20.37}$$

Choosing, e.g. the inter-plane distance $\delta = 1$ nm we obtain $T_c = O\left(10^2\right)$ K, showing that careful tuning of the optimal distance δ can lead to room-temperature superconductivity.

Chapter 21

Magnetic monopoles in loop-current Mott insulators

In Chapter 20 we have seen that magnetic monopoles can provide local seeds for strong-coupling electron pairing in higher angular momentum states. Here we will ask the question in which quantum materials can such monopoles appear. We have already mentioned effective monopoles due to curvature defects in graphene and graphite and to hedgehog defects in magnetically ordered materials, like cuprates. However, there is also another possibility, which we now discuss. To this end we shall consider Mott insulators, materials in which strong Coulomb interactions suppress charge transport. The main model to describe this physics is the Hubbard model, either in its fermionic or bosonic versions (for a review see [206]). The Hamiltonian H_{Hubbard} of this model is formulated on a lattice and contains two terms: a kinetic term with parameter t, describing the hopping of charges from one site to an adjacent one; and an on-site potential U modelling the Coulomb repulsion.

In the limit $U/t \to \infty$ charges are localized and the low-energy theory becomes a Heisenberg antiferromagnet for the spin degrees of freedom [206]. When U/t is large but finite, though, spontaneous circular loop currents can arise in the Hubbard model [207]. Moreover, loop currents have been proposed as the fundamental feature of the underdoped cuprates in their pseudogap state [208, 209], although only 2D planar currents have been considered, while the fundamental cell of the cuprates is 3D. Finally, loop-current order is emerging as a general feature of certain classes of Mott insulators [210] (for a general review of loop models see [211]).

Let us consider the simplest loop-current model on a cubic lattice with sites $\mathbf{x}$, lattice links i, and lattice step $\ell = 1$ (for simplicity of notation).

On the links of this lattice we introduce real loop currents $J_{\mathbf{x},i}$ such that $\hat{\Delta}_i J_{\mathbf{x},i} = 0$. In the framework of the Hubbard model, loop currents of $O(t^2 t'/U^2)$ around square plaquettes can be obtained, e.g., by allowing next-to-nearest neighbour hoppings with parameter t' and projecting on the gauge-invariant configuration (see below) where all next-to-nearest neighbour currents cancel out. Here, however, we abstract from the Hubbard model and we consider the simplest loop-current Hamiltonian

$$H = \sum_{\mathbf{x},i} \frac{1}{2f} J_{\mathbf{x},i}^2, \tag{21.1}$$

where f is a coupling constant. This describes a Mott insulating state in which percolating currents are suppressed by a large gap.

Loop currents generate magnetic fields. To be consistent, we have to add the corresponding energy to the Hamiltonian. We thus add time-independent gauge fields $A_{\mathbf{x},i}$ living on the links i. The corresponding magnetic fields live, in principle, on the links of the dual lattice. However, using the lattice Chern–Simons operator K_{ij} introduced in Chapter 14 one can assign the magnetic fields $B_{\mathbf{x},i} = K_{ij} A_{\mathbf{x},j}$ to the links of the primary lattice, which makes computations easier.

Since we are dealing with a lattice, there are two possible Hamiltonians describing the loop currents, the standard one,

$$H_{\mathrm{loop}} = \sum_{\mathbf{x},i} \frac{1}{2g} \left(\frac{1}{q} J_{\mathbf{x},i} - A_{\mathbf{x},i} \right)^2 + \frac{1}{2e^2} B_{\mathbf{x},i}^2, \tag{21.2}$$

(currents are unconstrained, longitudinal components can be compensated by gauge transformations, only the combination $(J/q - A)$ is gauge invariant) and the compact one

$$H_{\mathrm{compact}} = \sum_{\mathbf{x},i} \frac{1}{g} \left(1 - \cos \left(\frac{1}{q} J_{\mathbf{x},i} - A_{\mathbf{x},i} \right) \right) + \frac{1}{e^2} (1 - \cos B_{\mathbf{x},i}), \tag{21.3}$$

with q an integer and $g = f/q^2$. The two have the same local properties but differ in their topology, as we have already seen in previous chapters when considering the action of compact QED. Note that, as a consequence of the minimal coupling, compactness of the gauge fields implies also compactness of the loop currents. From now on we shall consider only *compact loop-current order*.

To proceed, let us use the Villain formulation (see Sec. 5.1) of the statistical partition function,

$$Z_{\text{compact}} = \sum_{\{m_{\mathbf{x},i}\}} \sum_{\{n_{\mathbf{x},i}\}} \int_{-\pi}^{+\pi} \mathcal{D}A_i \int_{-\pi q}^{\pi q} \mathcal{D}J_i$$

$$\times\, e^{-\beta \sum_{\mathbf{x},i} \left[\frac{1}{2g}\left(\frac{1}{q}J_{\mathbf{x},i} - A_{\mathbf{x},i} - 2\pi m_{\mathbf{x},i}\right)^2 - \frac{1}{2e^2}\left(B_{\mathbf{x},i} - 2\pi n_{\mathbf{x},i}\right)^2\right]},$$

$$(21.4)$$

where $\{m_{\mathbf{x},i}\}$ and $\{n_{\mathbf{x},i}\}$ are sets of integers defined on the links i at site $\mathbf{x}$ of the lattice and the lattice fields are integrated over their fundamental period. We now follow the treatment of compact QED in Sec. 5.3 by decomposing the integers $\{n_{\mathbf{x},i}\}$ as

$$n_{\mathbf{x},i} = K_{ij}k_{\mathbf{x},j} + K_{ij}\xi_{\mathbf{x},j} + \Delta_i\lambda_{\mathbf{x}}, \qquad (21.5)$$

where $\{k_{\mathbf{x},i}\}$ are integers and $\hat{\Delta}_i\Delta_i\lambda_{\mathbf{x}} = \nabla^2\lambda = m_{\mathbf{x}}$, also integer, representing magnetic monopoles. Note that both $\{\xi_{\mathbf{x},i}\}$ and $\{\lambda_{\mathbf{x}}\}$ are real but the combination $K_{ij}\xi_{\mathbf{x},j} + \Delta_i\lambda_{\mathbf{x}}$ describes longitudinal integers $\{l_{\mathbf{x},i}\}$ such that $\hat{\Delta}_i l_{\mathbf{x},i} = m_{\mathbf{x}}$, the Dirac strings. Therefore

$$\lambda_{\mathbf{x}} = \frac{\hat{\Delta}_i}{\nabla^2}l_{\mathbf{x},i}, \qquad (21.6)$$

and

$$\hat{K}_{ij}\frac{1}{\nabla^2}l_{\mathbf{x},j} + \xi_{\mathbf{x},j} = t_{\mathbf{x},i} + \Delta_i\phi_{\mathbf{x}}, \qquad (21.7)$$

with $\{t_{\mathbf{x},i}\}$ integers.

We now represent the integers $\{m_{\mathbf{x},i}\}$ as

$$m_{\mathbf{x},i} = \hat{K}_{ij}\frac{1}{\nabla^2}l_{\mathbf{x},j} + k_{\mathbf{x},j} + \xi_{\mathbf{x},j} - \Delta_i\phi_{\mathbf{x}} + r_{\mathbf{x},i}, \qquad (21.8)$$

where $\{r_{\mathbf{x},i}\}$ are integers.

When K_{ij} is applied on the first three terms of the integers $m_{\mathbf{x},j}$ one obtains the representation (21.5) of the integers $n_{\mathbf{x},i}$ up to an integer $l_{\mathbf{x},i}$ which can be reabsorbed anyhow into the set $\{r_{\mathbf{x},i}\}$. We now use the integers $\{k_{\mathbf{x},i}\}$ to shift the domain of the gauge fields $A_{\mathbf{x},i}$ from $[-\pi, +\pi]$ to $[-\infty, +\infty]$ and we absorb the real variables $\{\xi_{\mathbf{x},i}\}$ into these non-compact gauge fields. In the loop-current term of the Hamiltonian this is achieved by adding and subtracting $\{\xi_{\mathbf{x},i}\}$ and absorbing these two terms in $A_{\mathbf{x},i}$

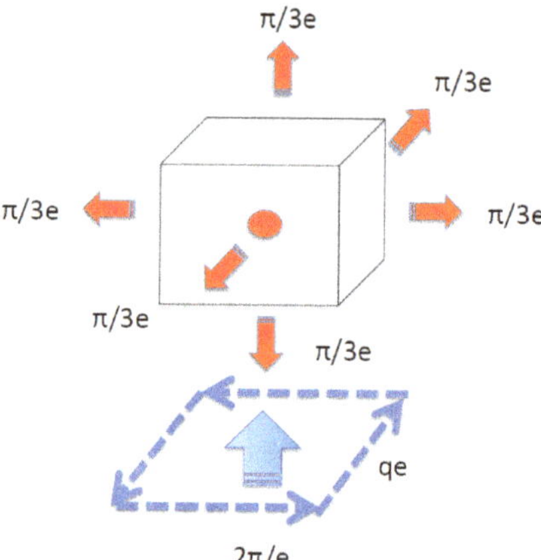

Fig. 21.1. An elementary loop-current monopole configuration. The magnetic fluxes out of the fundamental cube are shown in red. There are no loop currents on the edges of this cube since they all cancel out. The only loop currents are associated with the tower of plaquettes forming the Dirac string, sketched in blue. Because of the periodicity of both gauge fields and loop currents, the Dirac string does not carry any energy when a loop current of strength qe circulates around it.

and $J_{\mathbf{x},i}/q$ respectively, after also $J_{\mathbf{x},i}$ have been transformed into real variables, as explained below. We have thus replaced the set of three integers $\{n_{\mathbf{x},i}\}$ with a set of two integers $\{k_{\mathbf{x},i}\}$ (there are only two because of gauge invariance under transformations $k_{\mathbf{x},i} \to k_{\mathbf{x},i} + \Delta_i \chi_{\mathbf{x}}$) and one further set of integer $\{m_{\mathbf{x}}\}$, which represent magnetic monopoles.

For the integers $\{m_{\mathbf{x},i}\}$ we use the representation (21.7) and we absorb the integers $\{t_{\mathbf{x},i}\}$ into a redefinition of the integers $\{r_{\mathbf{x},i}\}$. These can then be used to shift the domain of the loop currents $J_{\mathbf{x},i}$ from $[-q\pi, +q\pi]$ to $[-\infty, +\infty]$. The partition function becomes thus

$$Z = \sum_{\{m_{\mathbf{x}}\}} \int_{-\infty}^{+\infty} \mathcal{D}A_i \int_{-\infty}^{+\infty} \mathcal{D}J_i \; \mathrm{e}^{-\beta(H_{\mathrm{loop}} + H_{\mathrm{monopoles}})},$$

$$H_{\mathrm{loop}} = \sum_{\mathbf{x},i} \left[\frac{1}{2g} \left(\frac{1}{q} J_{\mathbf{x},i} - A_{\mathbf{x},i} \right)^2 + \frac{1}{2e^2} B_{\mathbf{x},i}^2 \right], \qquad (21.9)$$

$$H_{\mathrm{monopoles}} = \frac{2\pi^2}{e^2} m_{\mathbf{x}} \frac{1}{-\nabla^2} m_{\mathbf{x}}.$$

Compact loop-current order admits thus magnetic monopoles interacting by the usual 3D Coulomb law. As we have derived in Chapter 14, these monopoles are gapped excitations for sufficiently low couplings. These gapped monopoles can be excited by doping the sample, since they can lower the overall energy by pairing the free charge carriers, as explained in Chapter 20. Loop-current insulators are thus candidates to become super-conductors upon doping via the creation of magnetic monopoles.

A magnetic flux through a plaquette of the lattice is equivalent, by the discrete Ampère law, to a loop current around the boundary of the plaquette. A magnetic monopole can thus, alternatively, be described by a specific loop current distribution. This is shown in Fig. 21.1. While a net magnetic flux $2\pi/e$ flows out of the cube containing the monopole, there are no loop currents around the edges of this cube, they all cancel out. Only the tower of plaquettes carrying the incoming magnetic flux supports loop currents. When a loop current of strength qe circulates around these plaquettes, however, this tube does not carry neither gauge nor matter energy and becomes a combined Dirac string.

Chapter 22

Interplay of disorder, topology, and interactions: Endogenous vs. exogenous disorder

As we have stressed repeatedly in previous chapters, in the superconductor-to-insulator (SIT) transition, exogenous disorder plays an important role as a tuning parameter, since it can influence the strength of the Coulomb interaction [95], but is irrelevant in the renormalization group sense, i.e. it does not influence the infrared character of the emerging phases. In this chapter we will focus on the interplay between disorder, topology, and interactions and explain in more detail why this is the case.

Since the seminal work of Anderson [31] it is known that disorder can localize single-particle states (for a review see [32]). This idea has then been generalized to interacting quantum states, although localization is, strictly speaking, established only in 1D [212]. This has led to the concept of many-body-localization (MBL) as a possible "glassy" phase of matter in which transport is suppressed, with an accompanying loss of ergodicity [213] due to an extensive number of local integrals of motion [214]. In this phase the system does not thermalize under its own dynamics since many of the possible states that are allowed on grounds of energy and momentum conservation are not anymore accessible by the time evolution.

While exogenous disorder is still considered as one of the main causes of transport suppression at finite temperatures in the condensed matter community, there is by now a vast body of literature showing that the same phenomena usually tied to MBL also arise in complete absence of disorder, mainly due to confinement [215–226], i.e. to interactions becoming extremely strong. The vast majority of recently studied examples are also 1D. The phenomenon, however, is generic and takes place in any number of dimensions. Its origin is confinement due to the condensation of topological defects, which play the role of an "endogenous", emergent disorder.

The theory of the condensation of topological defects in solid state media was originally developed by Julia and Toulouse in the 1970s [227]. They considered generic $(d+1)$-dimensional effective field theories with a symmetry group G spontaneously broken to H. The homotopy groups $\Pi_h(G/H)$ classify the topological defects that can arise in this theory (for a review see [228]). A non-trivial $\Pi_h(G/H)$ for $h < d$ describes solitons of dimension $d-h-1$; for $h = d$, instead, it describes instantons of the Euclidean version of the model (for a review see [229]). These finite-energy (action) classical solutions are characterized by radii $r_i = 1/M_i$, where M_i are the various masses associated with spontaneous symmetry breakdown. From the point of view of the low-energy effective theory with symmetry group H, valid on energy scales much smaller than $\min\{M_i\}$, they can be viewed essentially as singularities of dimension $d - h - 1$ in R^d (for solitons) or point singularities in R^{d+1} (for instantons).

Of course, the low-energy effective field theory with broken symmetry H is valid only "outside" of these singularities, in the interior of which the higher-energy degrees of freedom live. So far so well. But for certain parameters of the theory, be it temperature or a quantum coupling constant, the free energy of the topological defects may turn negative, in which case the system favours their proliferation. In this case the manifold on which the low-energy effective theory is defined soon becomes very complicated, representing an "endogenous disorder". The question arises as to what happens to the low-energy effective field theory when the topological defects condense. The Julia–Toulouse theory [227] provides an answer to this question: one can replace the complex manifold with proliferating singularities by the usual $\mathbb{R}^d$ provided one introduces new hydrodynamical modes describing the fluctuations of the continuous distribution of topological defects.

In general media, this procedure is not simple to implement. It was, however, shown that it becomes straightforward to realize in (generalized) Abelian gauge theories [160] involving antisymmetric tensor fields. This is important, since interactions in condensed matter can often be modeled as mediated by such generalized tensor gauge fields, the corresponding action encoding thus the response function of the system.

Consider in $(d+1)$-dimensional space-time a gauge theory for a $(h-1)$-form field a_{h-1} with compact gauge symmetry under transformations

$a_{h-1} \to a_{h-1} + d\lambda_{h-2}$ and action

$$S = \int \left(\frac{(-1)^{h-1}}{e^2} \, da_{h-1} \wedge (da_{h-1})^* + \, a_{h-1} \wedge j_{h-1}^* + S_M \right), \qquad (22.1)$$

where j_{h-1} describes a conserved (tensor) current of fields whose dynamics is governed by the action S_M. For convenience we will take the canonical mass dimension of a_{h-1} as $(d-1)/2$, so that e^2 is a dimensionless coupling constant. This action describes

$$N_{h-1} = \binom{d-1}{h-1} \qquad (22.2)$$

degrees of freedom.

When the gauge symmetry is compact, there exist $(d-h-1)$-dimensional singularities describing the topological defects. These are described by the form $f_h = da_{h-1}$ and are characterized by the homotopy groups $\Pi_h (G/H) = \mathbb{Z}$ encoding the topological invariants

$$\int_{S_h} f_h, \qquad (22.3)$$

where S_h is an h-dimensional sphere surrounding the singularity on a $(h+1)$-dimensional hyperplane perpendicular to it. When the gauge coupling e^2 is such that it favours the condensation of these topological defects, the Julia–Toulouse prescription [160, 227] states that one can replace this "endogenous disorder" by new hydrodynamical fields. These are obtained by promoting the form f_h to a new fundamental form with action

$$S_{\text{cond}} = \int \left[\frac{(-1)^h}{\Lambda^2} H_{h+1} \wedge H_{h+1}^* + \frac{(-1)^{h-1}}{e^2} (f_h - da_{h-1}) \wedge (f_h - da_{h-1})^* \right.$$

$$\left. + (f_h - da_{h-1}) \wedge T_h^* + S_M \right], \qquad (22.4)$$

where $H_{h+1} = df_h$, Λ is a new mass scale, and we will discuss the form T_h in a moment.

This action is invariant under the new, combined gauge symmetry

$$f_h \to f_h + d\psi_{h-1},$$
$$a_{h-1} \to a_{h-1} + \psi_{h-1}. \qquad (22.5)$$

This means that the original field a_{h-1} can be completely reabsorbed into f_h by gauge fixing. The gauge-fixed action becomes

$$S_{\text{cond}} = \int \left[\frac{(-1)^h}{\Lambda^2} H_{h+1} \wedge H_{h+1}^* + \frac{(-1)^{h-1}}{e^2} f_h \wedge f_h^* + f_h \wedge T_h^* + S_M \right],$$

$$(22.6)$$

and describes

$$N_{\text{cond}} = \binom{d}{h} \qquad (22.7)$$

massive degrees of freedom of mass $m = \Lambda/e$. This Julia–Toulouse mechanism is the dual of the usual Anderson–Higgs mechanism: the new hydrodynamical modes due to the condensation of defects, and encoded in the gauge-invariant components of f_h, have "eaten up" the original degrees of freedom, thereby acquiring mass. Note that there is no split up into phase and modulus of an order parameter, however. We are thus dealing with the dual of the original Stückelberg mechanism [230], rather than the Anderson–Higgs mechanism.

Let us now come to the most important point for present purposes. To match the promotion of the original massless $(h-1)$-form gauge field a_{h-1} to a massive h-form f_h we must also promote the original matter current j_{h-1} to a h-form T_h so that $dT_h^* = j^*$, i.e.

$$\partial_\mu T^{\mu\nu\ldots\alpha} = \frac{1}{h} j^{\nu\ldots\alpha}, \qquad (22.8)$$

so that the original model (22.1) is recovered in the limit $\Lambda \to 0$. This shows that the original matter currents j represent the boundaries of the new matter degrees of freedom. If the original matter degrees of freedom were point-like objects, a string develops between them; if they were closed rings, a membrane develops spanning the ring; if they were surfaces, a volume develops spanning the membrane and so on. The effective field theories for these new degrees of freedom can be obtained by integrating over the new gauge fields [160]. In all cases one obtains a (generalized) tension for the new geometric object spanning the original boundary excitations and these cannot be anymore separated on distances larger than a corresponding characteristic length, both determined by the scale Λ. The Julia–Toulouse phase due to the condensation of topological defects is a *confinement phase*. All these results can be obtained also by microscopic computations in lattice models [18, 141].

The "endogenous disorder" due to the condensation of topological defects completely prevents mobility and transport of the original matter degrees of freedom, which fall out of the spectrum of the infrared theory. A whole section of the Hilbert space which is allowed purely on momentum and energy conservation is not accessible anymore. The new "composite" degrees of freedom are typically much heavier than the original ones: their time evolution is thus quite slower. All these are features that are normally attributed to many-body localization by strong disorder. Confinement by the condensation of topological defects, however, leads to the same physics in complete absence of disorder. It also predates many-body localization by 30 years.

Let us now focus on the concrete example of the 2D SIT via an intermediate Bose metal state and let us review the picture of the SIT as derived in this book but with a spotlight on the role of disorder. When the strength of the Coulomb interaction is increased, either by decreasing the film thickness, turning on an orthogonal magnetic field, or increasing disorder, Cooper pairs are forced out of the condensate by strong quantum fluctuations. At this point, the infrared-dominant interaction is the topological interaction between out-of-condensate charges and dual vortices which, in its local formulation, amounts to a Chern–Simons term in the effective low-energy field theory. As a consequence, a topological ground state forms, which corresponds to a bosonic topological insulator. Depending only on the global, topological properties of the sample, this ground state is completely insensitive to disorder, which affects only local, geometric properties. Disorder can only localize the quasi-particles around this topological ground state which helps broaden it to a full-fledged intermediate phase, although it is not even necessary for this due to the formation of a Wigner crystal, as explained in Chapter 10. This is an exact parallel to the role of disorder in the fractional quantum Hall effect (FQHE). Topology dominates and leads to the formation of an incompressible quantum Hall fluid. Disorder affects only the fractionally charged quasi-particles of this topological fluid [231], although, also in this case, it is not instrumental for plateaus due to the formation of a Wigner crystal [94].

The intermediate state is characterized by a topological gap for bulk modes, quantified by the Chern–Simons mass. Near the superconducting transition this gap is very high so that only edge modes contribute to the conductance, leading to a pronounced metallic behaviour even at not extremely low temperatures. When the quantum coupling is increased, making Coulomb interactions stronger, the gap decreases. The metallic behaviour can be seen only at extremely low temperatures. At experimentally accessible temperatures, instead, bulk modes start to contribute, leading to an activated, insulating behaviour. In this region, localization of the bulk modes by exogenous disorder can indeed take place, not of the edge modes though. The important point, however, is that at a second critical coupling $g = g_{SI}$ the vortices become unstable due to the "condensation" of magnetic monopole instantons, an instance of the general Julia–Toulouse mechanism described above. At this critical point the electric interaction becomes linear and new, neutral extended states form, electric pions, exactly as what happens in the strong interaction. At this point all localization effects become subdominant with respect to the strong electric interaction, there are neither electrons nor Cooper pairs to localize anymore. In addition to the measured BKT universality class of this transition, the experimental detection of two kinks in the $I(V)$ curves at this transition (Fig. 18.2(b)) confirms this picture and seems to rule out MBL-based ideas for the onset of superinsulation [232–234]. The character of the phases at the SIT is dominated by topology and strong electric interactions.

Chapter 23

Synthesis

To conclude we would like to sum up in a compact form the main points of this book.

- Magnetic monopoles, while elusive as elementary particles, can exist as emergent excitations in condensed matter systems, both as instantons in 2D and as particles in 3D, where they can form a plasma (2D) or a Bose condensate (3D) which constitutes a superinsulating state with electric charge confinement and infinite resistance even at finite temperatures, the dual of a superconductor.
- Both in 2D and in 3D there exists a Bose metal which behaves as a Fermi liquid at low enough temperatures even if its constituents are bosonic Cooper pairs of charge $2e$. This behaviour arises because the bulk is frozen by topological interactions and transport arises exclusively from edge modes, typically along a Chalker–Coddington percolation network. In 2D this state is a bosonic topological insulator, in 3D it is the pseudogap state of high-T_c cuprates, which realizes a strong 3D superinsulator, a Bose condensate of dyons carrying both magnetic and electric charge.
- In the SIT disorder plays only the role of a tuning parameter affecting the strength of the Coulomb interaction but is irrelevant in the renormalization group sense, it does not influence the long-distance properties of the various possible phases around the SIT. These are determined by topology and the strength of the electric interaction, which becomes linear at a critical point where new, neutral extended states form.

- Magnetic monopoles, both effective ones and possibly also real ones, provide a pairing mechanism for real-space, localized electron pairs with binding energies of the order of the Fermi energy. By tuning the interplane distance in layered materials supporting these excitations, one can hope to obtain superconductors at room temperature.

Bibliography

[1] Y. Nambu, Strings, monopoles and gauge fields, *Phys. Rev.* **D10**, 4262 (1974).

[2] S. Mandelstam, Vortices and quark confinement in non-Abelian gauge theories, *Phys. Rep.* **23C**, 245–249 (1976).

[3] G. 't Hooft, On the phase transition towards permanent quark confinement, *Nucl. Phys.* **B138**, 1–25 (1978).

[4] J. Greensite, *An Introduction to the Confinement Problem*, Springer-Verlag, Berlin (2011).

[5] M. Tinkham, *Introduction to Superconductivity*, Dover Publications, New York (1996).

[6] M. C. Diamantini, P. Sodano, and C. A. Trugenberger, Gauge theories of Josephson junction arrays, *Nucl. Phys.* **B474**, 641–677 (1996).

[7] R. Fazio and H. van der Zant, Quantum phase transitions and vortex dynamics in superconducting networks, *Phys. Rep.* **355**, 235–334 (2001).

[8] K. B. Efetov, Phase transition in granulated superconductors, *Sov. Phys. JETP* **51**, 1015–1022 (1980); D. Haviland, Y. Liu, and A. Goldman, Onset of superconductivity in the two-dimensional limit, *Phys. Rev. Lett.* **62**, 2180–2183 (1989); A. Hebard and M. A. Paalanen, Magnetic-field-tuned superconductor-insulator transition in two-dimensional films, *Phys. Rev. Lett.* **65**, 927–930 (1990); R. Fazio and G. Schön, Charge and vortex dynamics in arrays of tunnel junctions, *Phys. Rev.* **B43**, 5307–5320 (1991).

[9] M. P. A. Fisher and G. Grinstein, Quantum critical phenomena in charged superconductors, *Phys. Rev. Lett.* **60**, 208–211 (1988); M. P. A. Fisher, Quantum phase transitions in disordered two-dimensional superconductors, *Phys. Rev. Lett.* **65**, 923–926 (1990); M. P. A. Fisher, G. Grinstein, and S. M. Girvin, Presence of quantum diffusion in two dimensions: Universal resistance at the superconductor–insulator transition, *Phys. Rev. Lett.* **64**, 587–590 (1990); S. M. Girvin, M. Wallin, M.-C. Cha, M. P. A. Fisher and P. Young, Universal conductivity as the superconductor–insulator transition in two dimensions, *Prog. Theor. Phys. Supp.* **107**, 135–144 (1992).

[10] A. Krämer and S. Doniach, *Phys. Rev. Lett.* **81**, 3523 (1998).

[11] V. M. Vinokur *et al.*, Superinsulator and quantum synchronization, *Nature* **452**, 613–615 (2008).

[12] T. I. Baturina and V. M. Vinokur, Superinsulator–superconductor duality in two dimensions, *Ann. Phys.* **331**, 236–257 (2013).

[13] G. Sambandamurthy, L. W. Engels, A. Johansson, E. Peled and D. Shahar, Experimental evidence for a collective insulating state in two-dimensional superconductors, *Phys. Rev. Lett.* **94**, 017003 (2005).

[14] V. L. Berezinskii, Destruction of long-range order in one-dimensional and two-dimensional systems having a continuous symmetry group I. Classical systems, *Sov. Phys. JETP* **32**, 493–500 (1970); J. M. Kosterlitz and D. J. Thouless, Long-range order and metastability in two-dimensional solids and superfluids: Application of dislocation theory, *J. Phys.* **C5**, L124 (1972); J. M. Kosterlitz and D. J. Thouless, Ordering, metastability and phase transitions in two-dimensional systems, *J. Phys.* **C6**, 1181–1203 (1973).

[15] A. Yu. Mironov, D. M. Silevitch, T. Proslier, S. V. Postolova, M. V. Burdastyh, A. K. Gutakovskii, T. F. Rosenbaum, V. M. Vinokur, and T. I. Baturina, Charge Berezinskii–Kosterlitz–Thouless transition in superconducting NbTiN films, *Sci. Rep.* **8**, 4082 (2018).

[16] M. C. Diamantini, C. A. Trugenberger, V. M. Vinokur, Confinement and asymptotic freedom with Cooper pairs, *Comm. Phys.* **1**(77), 1 (2018).

[17] P. Goddard and D. I. Olive, Magnetic monopoles in gauge field theories, *Rep. Prog. Phys.* **41**, 91 (1978).

[18] A. M. Polyakov, Compact gauge fields and the infrared catastrophe, *Phys. Lett.* **59**, 82–84 (1975).

[19] A. M. Polyakov, *Gauge Fields and Strings*, Harwood Academic Publisher, Chur, Switzerland (1987).

[20] S. Coleman, *Aspects of Symmetry*, Cambridge University Press, Cambridge (1985).

[21] C. A Trugenberger, M. C. Diamantini, N. Poccia, F. S. Nogueira, and V. M. Vinokur, Magnetic monopoles and superinsulation in Josephson junction arrays, *Quant. Rep.* **2**, 388–399 (2020).

[22] M. C. Diamantini, L. Gammaitoni, C. Strunk, S. V. Postolova, A. Yu. Mironov, C. A. Trugenberger, and V. M. Vinokur, Direct probe of the interior of an electric pion in a Cooper pair superinsulator, *Nat. Comm. Phys.* **3**, 142 (2020).

[23] C. Schmidt and S. Sharma, The phase structure of QCD, *J. Phys.* **G44**, 104002 (2017).

[24] D. Gross, Twenty five years of asymptotic freedom, *Nucl. Phys.* **B74**, 426–446 (1998).

[25] D. Das and S. Doniach, Existence of a Bose metal at $T = 0$, *Phys. Rev.* **B60**, 1261–1275 (1999); D. Das and S. Doniach, Bose metal: Gauge field fluctuations and scaling for field-tuned quantum phase transitions, *Phys. Rev.* **B64**, 134511 (2001).

[26] D. Dalidovich and P. Phillips, Phase glass is a Bose metal: A new conducting state in two dimensions, *Phys. Rev. Lett.* **89**, 27001 (2002); P. Pillips and D. Dalidovich, The elusive Bose metal, *Science* **302**, 243–247 (2003).

[27] A. Kapitulnik, S. A. Kivelson, and B. Spivak, Anomalous metals: Failed superconductors, *Rev. Mod. Phys.* **91**, 011002 (2019).

[28] S. Eley, S. Gopalakrishnan, P. M. Goldbart, and N. N. Mason, Approaching zero-temperature metallic states in mesoscopic superconductor–normal–superconductor array, *Nat. Phys.* **8**, 59–62 (2012).

[29] Z. Han *et al.*, Collapse of superconductivity in a hybrid tin-graphene Josephson junction array, *Nat. Phys.* **10**, 380–386 (2014).

[30] C. G. L. Bottcher *et al.*, Superconducting, insulating and anomalous metallic regimes in a gated two-dimensional semiconductor–superconductor array, *Nat. Phys.* **14**, 1138–1144 (2018).

[31] P. Anderson, Absence of diffusion in certain random lattices, *Phys. Rev.* **109**, 1492–1505 (1958).

[32] A. Abrahams (ed.), *Fifty Years of Anderson Localization*, World Scientific, Singapore (2010).

[33] R. Shankar, Topological insulators: A review, arXiv:1804.06471 (2018).

[34] M. C. Diamantini, A. Yu. Mironov, S. V. Postolova, X. Liu, Z. Hao, D. M. Silevitch, Ya. Kopelevich, P. Kim, C. A. Trugenberger, and V. M. Vinokur, Bosonic topological intermediate state in the superconductor–insulator transition, *Phys. Lett.* **A384**, 126570 (2020).

[35] D. Kowal and Z. Ovadyahu, Disorder induced granularity in an amorphous superconductor, *Solid St. Comm.* **90**, 783–786 (1994); M. V. Fistul, V. M. Vinokur, and T. I. Baturina, Collective Cooper-pair transport in the insulating state of Josephson-junction arrays, *Phys. Rev. Lett.* **100**, 086805 (2008).

[36] M. C. Diamantini, C. A. Trugenberger, and V. M. Vinokur, Quantum magnetic monopole condensate, *Nat. Comm. Phys.* **4**, 25 (2021).

[37] C. Parra, F. Niemstemski, A. W. Contryman, P. Giraldo-Gallo, T. H. Geballe, I. R. Fisher, and H. C. Manoharan, Signatures of two-dimensional superconductivity emerging within a three-dimensional host superconductor, *PNAS* **118**, e2017810118 (2021).

[38] C. Castelnovo, R. Moessner, and S. L. Sondhi, Magnetic monopoles in spin ice, *Nature* **451**, 42 (2008).

[39] X.-L. Qi, R. Li, J. Zang, and S.-C. Zhang, Inducing a magnetic monopole with topological surface states, *Science* **323**, 1184 (2009).

[40] S. B. Treiman, R. Jackiw, B. Zumino, and E. Witten, *Current Algebra and Anomalies* World Scientific, Singapore (1985).

[41] M. C. Diamantini, C. A. Trugenberger, and V. M. Vinokur, Topological nature of high-temperature superconductivity, *Adv. Quant. Tech.* **4**, 2000135 (2021).

[42] C. Proust and L. Taillefer, The remarkable underlying ground states of cuprate superconductors, *Ann. Rev. Condens. Matter Phys.* **10**, 409–429 (2019).

[43] M. C. Diamantini, C. A. Trugenberger, and V. M. Vinokur, Pseudo magnetic monopole mechanism for localized electron pairing in HTS, arXiv:2102:08652 (2021).

[44] Y. Aharonov and D. Bohm, Significance of electromagnetic potentials in quantum theory, *Phys. Rev.* **115**, 485–491 (1961).

[45] P. A. M. Dirac, Quantized singularities in the electromagnetic field, *Proc. Roy. Soc. (London)* **A133**, 60 (1931).

[46] M. C. Diamantini and C. A. Trugenberger, Superinsulators, a toy realization of confinement in condensed matter physics, in *Festschrift for Roman Jackiw's 80th birthday*, World Scientific, Singapore (2020).

[47] P. Minnhagen, The two-dimensional Coulomb gas, vortex unbinding and superfluid–superconducting films, *Rev. Mod. Phys.* **59**, 1001–1066 (1987).

[48] R. Jackiw and S. Templeton, How super-renormalizable interactions cure infrared divergences, *Phys. Rev.* **D23**, 2291 (1981); S. Deser, R. Jackiw, and S. Templeton, Three-dimensional massive gauge theories, *Phys. Rev. Lett.* **48**, 975 (1982); S. Deser, R. Jackiw, and S. Templeton, Topologically massive gauge theories, *Ann. Phys.* (N.Y.) **140**, 372–411 (1982).

[49] S. Weinberg, *Gravitation and Cosmology*, John Wiley & Sons, New York (1972).

[50] P. W. Anderson, Plasmons, gauge invariance and mass, *Phys. Rev.* **130**, 439–442 (1963).

[51] P. W. Higgs, Broken symmetries and the masses of gauge bosons, *Phys. Rev. Lett.* **13**, 508–509 (1964).

[52] E. P. Wigner, On unitary representations of the inhomogenous Lorentz group, *Ann. Math.* **40**, 149–204 (1939).

[53] L. H. Kaufmann, *Formal Knot Theory*, Princeton University Press, Princeton (1983).

[54] Y. Aharonov and A. Casher, Topological quantum effects for neutral particles, *Phys. Rev. Lett.* **53**, 319–321 (1984).

[55] F. Wilczek, Disassembling anyons, *Phys. Rev. Lett.* **69**, 132–135 (1992).

[56] G. Dunne, R. Jackiw, and C. A. Trugenberger, Topological (Chern–Simons) quantum mechanics, *Phys. Rev.* **D41**, 661–666 (1990).

[57] H. Kleinert, *Gauge Fields in Condensed Matter*, Vol. 1: Superflow and Vortex Lines (Disorder Fields, Phase Transitions), World Scientific, Singapore (1989).

[58] N. D. Mermin and H. Wagner, Absence of ferromagnetism or antiferromagentism in one- or two -dimensional isotropic Heisenberg models, *Phys. Rev. Lett.* **17**, 1133–1136 (1966); P. C. Hohenberg, Existence of long-range order in one and two dimensions, *Phys. Rev.* **152**, 383 (1967).

[59] J. Zinn-Justin, *Quantum Field Theory and Critical Phenomena*, Oxford University Press, Oxford (2002).

[60] S. Datta, *Electronic Transport in Mesoscopic Systems*, Cambridge University Press, Cambridge (1995).

[61] A. D. Zaikin, D. S. Golubev, A. van Otterlo, and G. T. Zimány, Quantum phase slips and transport in ultrathin superconducting wires, *Phys. Rev. Lett.* **78**, 1552–1555 (1997).

[62] M.-S. Choi, Y. J. Choi, M. Y. Choi, and S.-I. Lee, Quantum phase transitions in Josephson junction chains, *Phys. Rev.* **B57**, R716–R719 (1998).

[63] K. A. Matveev, A. I. Larkin, and L. I. Glazman, Persistent current in superconducting nanorings, *Phys. Rev. Lett.* **89**, 096802 (2002).

[64] H. P. Büchler, V. B. Geshkenbein, and G. Blatter, Quantum fluctuations in thin superconducting wires of finite length, *Phys. Rev. Lett.* **92**, 067007 (2004).

[65] M. B. Einhorn and R. Savit, Phase transitions and confinement in the Abelian Higgs model, *Phys. Rev.* **D19**, 1198 (1979).

[66] M. Caselle, M. Panero, and D. Vadacchino, Width of the flux tube in compact U(1) gauge theory in three dimensions, *JHEP* **02**, 180 (2016).

[67] J. C. Toledano and P. Toledano, *The Landau Theory of Phase Transitions*, World Scientific, Singapore (1987).

[68] S. Sachdev, *Quantum Phase Transitions*, Cambridge University Press, Cambridge (2011).

[69] R. B. Laughlin, Primitive and composite ground states in the fractional quantum Hall effect, *Surface Sci.* **142**, 163–172 (1984).

[70] K. V. Klizting, G. Dorda, and M. Pepper, New method for high-accuracy determination of the fine-structure constant based on quantized Hall resistance, *Phys. Rev. Lett.* **45**, 494–497 (1980).

[71] D. C. Tsui, H. L. Stormer, and A. C. Gossard, Two-dimensional magnetotransport in the extreme quantum limit, *Phys. Rev. Lett.* **48**, 1559 (1982).

[72] R. B. Laughlin, Anomalous quantum Hall effect: An incompressible quantum fluid approach, *Phys. Rev. Lett.* **50**, 1395–1398 (1983).

[73] P. Di Francesco, P. Mathieu, and D. Sénéchal, *Conformal Field Theory*, Springer Verlag, New York (1997).

[74] A. Cappelli, C. A. Trugenberger, and G. Zemba, Infinite symmetry in the quantum Hall effect, *Nucl. Phys.* **B396**, 465–490 (1993).

[75] A. Cappelli, C. A. Trugenberger, and G. Zemba, Stable hierarchical quantum Hall fluids as $W_{1+\infty}$ minimal models, *Nucl. Phys.* **B448**, 470 (1995).

[76] X. G. Wen and A. Zee, Classification of Abelian quantum Hall states and matrix formulation of topological fluids, *Phys. Rev.* **B46**, 2290 (1992).

[77] X.-G. Wen, Topological orders in rigid states, *Int. J. Mod. Phys.* **B4**, 239 (1990).

[78] M. Levin and X.-G. Wen, String-net condensation: A physical mechanism for topological phases, *Phys. Rev.* **B71**, 045110 (2005).

[79] F. Wilczek, *Fractional Statistics and Anyon Superconductivity*, World Scientific, Singapore (1990).

[80] X.-G. Wen, Topological order: From long-range entangled quantum matter to a unified origin of light and electrons, *ISRN* **2013**, 198710 (2013).

[81] R. Floreanini and R. Jackiw, Self-dual fields as charge-density solitons, *Phys. Rev. Lett.* **59**, 1873–1876 (1987).

[82] X.-G. Wen, Theory of the edge states in fractional quantum Hall effects, *Int. J. Mod. Phys.* **B6**, 1711–1762 (1992).

[83] N. O. Agasyan and K. Zarembo, Phase structure and nonperturbative states in three-dimensional adjoint Higgs model, *Phys. Rev.* **D57**, 2475 (1998).

[84] C. Itzykson and J.-M. Drouffe, *Statistical Field Theory*, Cambridge University Press, Cambridge (1989).

[85] A. V. Chaplik and M. V. Entin, Charged impurities in very thin layers, *Zh. Eksp. Teor. Fiz.* **61**, 2496–2503 (1971).

[86] L. V. Keldysh, Coulomb interaction in thin semiconductor and semimetal films, *JETP Lett.* **29**, 658–661 (1979).

[87] E. Dagotto, Complexity in strongly correlated electronic systems, *Science* **309**, 257–262 (2005).

[88] B. Sacépé *et al.*, Disorder-induced inhomogeneities of the superconducting state close to the superconductor–insulator transition, *Phys. Rev. Lett.* **101**, 157006 (2008).

[89] B. Sacépé *et al.*, Localization of preformed Cooper pairs in disordered superconductors, *Nat. Phys.* **7**, 239–244 (2011).

[90] M. Strongin, R. S. Thomson, O.F. Kammerer, and J. E. Crow, Destruction of superconductivity in disordered near-monolayer films, *Phys. Rev.* **B1**, 1078–1091 (1970); J. E. Crow and M. Strongin, Influence of normal-state electrical resistance on the superconducting transition temperature of ultrathin films, *Phys. Rev.* **3**, 2365 (1971).

[91] N. Trivedi, R. T. Scalettar, and M. Randeria, Superconductor–insulator transition in a disordered electronic system, *Phys. Rev.* **B54**, R3756 (1996).

[92] B. Sacépé, M. Feigel'man, and T. M. Klapwijk, Quantum breakdown of superconductivity in low-dimensional materials, *Nat. Phys.* **16**, 734–746 (2020).

[93] R. E. Prange and S. M. Girvin (eds.), *The Quantum Hall Effect*, Springer Verlag, Berlin (1987).

[94] K.-S. Kim and S. A. Kivelson, The quantum Hall effect in the absence of disorder, *npj Quantum Mater.* **6**, 22 (2021).

[95] A. M. Finkel'shtein, Superconducting transition temperature in amorphous films, *JETP Lett.* **45**, 46–49 (1987).

[96] T. Banks, R. Myerson, and J. Kogut, Phase transitions in Abelian lattice gauge theories, *Nucl. Phys.* **B129**, 493–510 (1977).

[97] Y.-M. Lu and A. Vishwanath, Theory and classification of interacting integer topological phases in two dimensions: A Chern–Simons approach, *Phys. Rev.* **B86**, 125119 (2012); C. Wang and T. Senthil, Boson topological insulators: A window into highly entangled quantum phases, *Phys. Rev.* **B87**, 235122 (2013); X. Chen, Z.-C. Gu, Z.-X. Liu, and X.-G. Wen, Symmetry protected topological orders and the group cohomology of their symmetry group, *Phys. Rev.* **B87**, 155114 (2013).

[98] Y. Arutyunov, D. S. Golubev, and A. D. Zaikin, Superconductivity in one dimension, *Phys. Rep.* **464**, 1–70 (2008).

[99] M. Z. Hasan and C. L. Kane, Topological insulators, *Rev. Mod. Phys.* **82**, 3045 (2010).

[100] X. Chen, Z.-C. Gu, Z.-X. Liu and X.-G. Wen, Symmetry protected topological orders and the group cohomology of their symmetry group, *Phys. Rev.* **B87**, 155114 (2013).

[101] S. Murakami, Two-dimensional topological insulators and their edge states, *J. Phys. Conf. Ser.* **302**, 012019 (2011).

[102] K. Wang, D. Graf, L. Li, L. Wang, and C. Petrovic, Anisotropic giant magnetoresistance in Nb_2Sb_2, *Sci. Rep.* **4**, 7328 (2015).

[103] G. Dunne, R. Jackiw, and C. A. Trugenberger, Chern–Simons theory in the Schrödinger representation, *Ann. Phys.* **194**, 197–223 (1989).

[104] M. C. Diamantini, C. A. Trugenberger, and V. Vinokur, Superconductor-to-insulator transition in absence of disorder, *Phys. Rev.* **B103**, 174516 (2021).

[105] M. C. Diamantini and C. A. Trugenberger, Higgsless superconductivity from topological defects in compact BF terms, *Nucl. Phys.* **891**, 401–419 (2015).

[106] E. B. Sonin, The Magnus force in superfluids and superconductors, *Phys. Rev.* **B55**, 485 (1996).

[107] D. Yoshioka, A. MacDonald, and S. Girvin, Fractional quantum Hall effect in two-layered systems, *Phys. Rev.* **B39** 1932 (1989).

[108] T. Senthil and M. Levin, Integer quantum hall effect for bosons, *Phys. Rev. Lett.* **110**, 046801 (2013).

[109] Y. P. Monarkha and V. E. Syvokon, A two-dimensional Wigner crystal, *Low Temp. Phys.* **38**, 1067 (2012).

[110] C. Xu and T. Senthil, Wave functions of bosonic symmetry protected topological phases, *Phys. Rev.* **B87**, 174412 (2013).

[111] J. M. Kosterlitz, The critical properties of the two-dimensional XY model, *J. Phys.* **C7**, 1046 (1974).

[112] A. B. Harris, Upper bounds for the transition temperatures of generalized Ising models, *J. Phys.* **C7**, 1671 (1974).

[113] J. T. Chalker and P. D. Coddington, Percolation, quantum tunneling and the integer Hall effect, *J. Phys.* **C21**, 2665 (1988).

[114] T. Ochiai, Gapless surface states in a three-dimensional Chalker–Coddington type network model, arXiv:1510.04033 (2015).

[115] R. Fazio and G. Schön, Charge and vortex dynamics in arrays of tunnel junctions, *Phys. Rev.* **B43**, 5307–5320 (1991).

[116] A. van Otterlo, R. Fazio, and G. Schön, Quantum vortex dynamics in Josephson junction arrays, *Physica* **B203**, 504–512 (1994).

[117] H. van der Zant, F. C. Fritschy, T. P. Orlando, and J. E. Mooji, Ballistic motion of vortices in Josephson junction arrays, *Europhys. Lett.* **18**, 343–512 (1992).

[118] H. S. J. van der Zant, F. C. Fritschy, W. J. Elion, L. J. Geerligs, and J. E. Mooij, Field-induced superconductor-to-insulator transitions in Josephson junction arrays, *Phys. Rev. Lett.* **69**, 2971–2974 (1992).

[119] N. Poccia, T. I. Baturina, F. Coneri, C. G. Molenaar, X. R. Wang, G. Bianconi, A. Brinkman, H. Hilgenkamp, A. A. Golubov, and V. M. Vinokur, Critical behavior at a dynamic vortex insulator-to-metal transition, *Science* **349**, 1202–1205 (2015).

[120] A. A. Belavin, A. M. Polyakov, and A. B. Zamolodchikov, Infinite conformal symmetry in two-dimensional quantum field theory, *Nucl. Phys.* **B241**, 333 (1984).

[121] M. C. Diamantini and C. A. Trugenberger, Minimal models for a superconductor–insulator conformal quantum phase transition, *J. Phys.* **A46**, 115003 (2013).

[122] E. Ardonne, P. Fendley, and E. Fradkin, Topological order and conformal quantum critical points, *Ann. Phys.* **310**, 493–551 (2004).

[123] H. S. J. van der Zant, W. J. Elion, L. J. Geerligs, and J. E. Mooij, *Phys. Rev.* **B54**, 10081 (1996).

[124] J. K. Jain, Microscopic theory of the fractional quantum Hall effect, *Adv. Phys.* **44**, 105 (1992); J. K. Jain, Composite fermions in the quantum Hall regime, *Science* **266**, 1199 (1994).

[125] G. Cristofano, G. Maiella, R. Musto, and F. Nicodemi, Topological order in quantum Hall effect and two-dimensional conformal field theory, *Nucl. Phys. Proc. Supp.* **B33**, 119–133 (1993).

[126] J.-B. Zuber, Conformal field theories, Coulomb gas picture, integrable models, in *Fields, Strings, Critical Phenomena*, Proceedings of Les Houches Summer School 1988, E. Brézin and J. Zinn-Justin (eds.), North-Holland, Amsterdam (1990).

[127] B. Wybourne, *Classical Groups for Physicists*, John Wiley & Sons, New York (1974).

[128] X. Shen, $W\infty$ and string theory, *Int. J. Mod. Phys.* **A7**, 6953 (1990).

[129] C. A. Trugenberger, 2D superconductivity: Classification of universality classes by infinite symmetry, *Nucl. Phys.* **B176**, 509–518 (2005).

[130] V. Kac and A. Radul, Quasifinite highest weight modules over the Lie algebra of differential operators on the circle, *Comm. Math. Phys.* **157**, 429–457 (1993).

[131] V. A. Fateev and A. B. Zamolodchikov, Conformal quantum field theory models in two dimensions having Z_3 symmetry, *Nucl. Phys.* **B280**, 644 (1987); V. A. Fateev and S. L. Lykyanov, The models of two-dimensional conformal quantum field theory with Z_n symmetry, *Int. J. Mod. Phys.* **A3**, 507 (1988).

[132] G. Blatter, M. V. Feigelman, V. B. Geshkenbein, A. I. Larkin, and V. M. Vinokur, Vortices in high-temperature superconductors, *Rev. Mod. Phys.* **66**, 1125–1388 (1994).

[133] M. Kalb and P. Ramond, Classical direct interstring action, *Phys. Rev.* **D9**, 2273–2284 (1974).

[134] D. Birmingham, M. Blau, M. Rakowski, and G. Thompson, Topological field theory, *Phys. Rep.* **209**, 129 (1991).

[135] M. Bergeron, G. W. Semenoff, and R. Szabo, Canonical BF-type topological field theory and fractional statistics of strings, *Nucl. Phys.* **B437**, 695 (1995).

[136] A. Vishwanath and T. Senthil, Physics of three-dimensional bosonic topological insulators: Surface-deconfined criticality and quantized magneto-electric effect, *Phys. Rev.* **X3**, 011016 (2013).

[137] T. Allen, M. Bowick, and A. Lahiri, *Mod. Phys. Lett.* **A6**, 559 (1991).

[138] D. Nelson, T. Piran, and S. Weinberg, *Statistical Mechanics of Membranes and Surfaces*, World Scientific, Singapore (2004).

[139] M. C. Diamantini and C. A. Trugenberger, Surfaces with long-range correlations from non-critical strings, *Phys. Lett.* **B421**, 196–202 (1998); M. C. Diamantini and C. A. Trugenberger, Geometric aspects of confining strings, *Nucl. Phys.* **B531**, 151–167 (1998); M. C. Diamantini, H. Kleinert, and C. A. Trugenberger, Strings with negative stiffness and hyperfine structure, *Phys. Rev. Lett.* **82**, 267–270 (1999).

[140] M. C. Diamantini, P. Sodano, and C. A. Trugenberger, Superconductors with topological order, *Eur. Phys. J.* **B53**, 19 (2016).

[141] P. Orland, Instantons and disorder in antisymmetric tensor gauge fields, *Nucl. Phys.* **B205**, 107–118 (1982).

[142] C. Bonati and M. D'Elia, Phase diagram of the 4D U(1) model at finite temperature, *Phys. Rev.* **D88**, 065025 (2013).

[143] M. C. Diamantini, L. Gammaitoni, C. A. Trugenberger, and V. M. Vinokur, Vogel–Fulcher–Tamman criticality of 3D superinsulators, *Sci. Rep.* **8**, 15718 (2018).

[144] P. W. Anderson, Lectures on amorphous systems, in *Les Houches*, Session XXXI, R. Balian *et al.* (eds.), North-Holland, Amsterdam (1978).

[145] M. Ovadia *et al.*, Evidence for a finite-temperature insulator, *Sci. Rep.* **5**, 13503 (2015).

[146] M. G. Vasin, V. N. Ryzhov, and V. M. Vinokur, Berezinskii–Kosterlitz–Thouless and Vogel–Fulcher–Tammann criticality in the XY model, arXiv:1712.00757 (2017).

[147] A. J. Niemi and G. Semenoff, Axial anomaly anduced fermion fractionization and effective gauge theory actions in odd-dimensional space-times, *Phys. Rev. Lett.* **51**, 2077 (1983).

[148] A. N. Redlich, Gauge non-invariance and parity non-conservation of three-dimensional fermions, *Phys. Rev. Lett.* **52**, 18 (1984); A. N. Redlich, Parity violation and gauge non-invariance of the effective gauge field action in three dimensions, *Phys. Rev.* **D29**, 2366 (1984).

[149] I. D. Choudury, A. Lahiri, M. C. Diamantini, G. Guarnaccia, and C. A. Trugenberger, 4D topological mass by gauging spin, *JHEP* **1506**, 081 (2015).

[150] S. Weinberg, *The Quantum Theory of Fields*, Cambridge University Press, Cambridge (1995).

[151] A. S. Golhaber and D. Wilkinson, Composite magnetic monopoles for SU(N) spontaneously broken to U(1), *Nucl. Phys.* **B114**, 317–333 (1976); R. Jackiw and C. Rebbi, Spin from isosopin in a gauge theory, *Phys. Rev. Lett.* **36**, 1116 (1976); P. Hasenfratz and G. 't Hooft, A fermion–boson puzzle in a gauge theory, *Phys. Rev. Lett.* **36**, 1119 (1976).

[152] T. T. Wu and C. N. Yang, Dirac monopole without strings: Monopole harmonics, *Nucl. Phys.* **B107**, 365–380 (1976).

[153] A. S. Goldhaber, R. MacKenzie, and F. Wilczek, Field corrections to induced statistics, *Mod. Phys. Lett.* **A4**, 21–31 (1989).

[154] F. Wilczek, Two applications of axion electrodynamics, *Phys. Rev. Lett.* **58**, 1799 (1987).

[155] D. Lyons, An elementary introduction to the Hopf fibration, *Mathematics Magazine* **76**, 87–98 (2003).

[156] E. Witten, Dyons of charge $\theta/2\pi$, *Phys. Lett.* **86**, 283–287 (1979).

[157] A. M. Essin, J. E. Moore, and D. Vanderbilt, Magnetoelectric polarizability and axion electrodynamics in crystalline insulators, *Phys. Rev. Lett.* **102**, 146805 (2009).

[158] L. Wu, M. Salehi, N. Koirala, J. Moon, S. Oh, and N. P. Armitage, Quantized Faraday and Kerr rotation and axion electrodynamics of a 3D topological insulator, *Science* **254**, 1124–1127 (2016).

[159] M. C. Diamantini, F. Quevedo, and C. A. Trugenberger, Confining strings with topological term, *Phys. Lett.* **B396**, 115–121 (1997).

[160] F. Quevedo and C. A. Trugenberger, Phases of antisymmetric tensor field theories, *Nucl. Phys.* **B501**, 143–172 (1997).

[161] A. Polyakov, Confining strings, *Nucl. Phys.* **B486**, 23–33 (1997).

[162] J. Polchinski, *String Theory*, Cambridge University Press, Cambridge (1998).

[163] A. Polyakov, Fine structure of strings, *Nucl. Phys.* **B268**, 406–412 (1986).

[164] A. Polyakov, Two-dimensional quantum gravity, in *Fields, Strings and Critical Phenomena*, Les Houches session XLIX, 1998, North-Holland, Amsterdam (1990).

[165] F. S. Nogueira and H. Kleinert, Compact quantum electrodynamics in 2+1 dimensions and spinon confinement: A renormalization group analysis, *Phys. Rev.* **B77**, 045107 (2008); H. Kleinert, F. Nogueira, and A. Sudbo, Kosterlitz–Thouless-like deconfinement mechanism in the (2+1)-dimensional Abelian Higgs model, *Nucl. Phys.* **B666**, 361–395 (2003).

[166] S. Nagy, Lectures on renormalization and asymptotic safety, *Ann. Phys.* **350**, 310–346 (2014).

[167] M. Lüscher, Symmetry-breaking aspects of the roughening transition in gauge theories, *Nucl. Phys.* **B180**, 317–329 (1981).

[168] R. S. Newrock, C. J. Lobb, U. Geigenmüller, and M. Octavio, The two-dimensional physics of Josephson junction arrays, *Solid State Phys.* **54**, 263 (2000).

[169] N. Barišić *et al.*, Universal sheet resistance and revised phase diagram of the cuprate high-temperature superconductors, *PNAS* **110**, 12235–12240 (2013).

[170] C. Proust, B. Vignolle, J. Levallois, S. Adachi, and N. E. Hussey, Fermi liquid behavior of the in-plane resistivity in the pseudogap state of $YBa_2Cu_4O_8$, *PNAS* **113**, 13564–13659 (2016).

[171] P. Zhou *et al.*, Electron pairing in the pseudogap state revealed by shot noise in copper oxide junctions, *Nature* **572**, 493–496 (2019).

[172] B. Keimer, S. A. Kivelson, M. R. Norman, S. Uchida, and J. Zaanen, From quantum matter to high-temperature superconductivity in copper oxides, *Nature* **518**, 179–186 (2015).

[173] Y. Yu *et al.*, High-temperature superconductivity in monolayer $Bi_2Sr_2CaCu_2O_{8+\delta}$, *Nature* **575**, 156–163 (2019).

[174] S. Mukhopadhyay *et al.*, Evidence for a vestigial nematic state in the cuprate pseudogap phase, *PNAS* **116**, 13249–13254 (2019).

[175] Y. Sato *et al.*, Thermodynamic evidence for a nematic phase transition at the onset of the pseudogap in $YBa_2Cu_3O_y$, *Nat. Phys.* **13**, 1074–1078 (2017).

[176] S. A. Kivelson, E. Fradkin, and V. J. Emery, Electronic liquid-crystal phases of a doped Mott insulator, *Nature* **393**, 550–553 (1998).

[177] L. Nie, L. Tarjus, and S. A. Kivelson, Quenched disorder and vestigial nematicity in the pseudogap regime of the cuprates, *PNAS* **111**, 7980–7985 (2014).

[178] A. Legros *et al.*, Universal T-linear resistivity and Planckian dissipation in overdoped cuprates, *Nat. Phys.* **15**, 142–147 (2019).

[179] P. Giraldo-Gallo *et al.*, Scale-invariant magnetoresistance in a cuprate superconductor, *Science* **361**, 479–481 (2018).

[180] G. Q. Zheng, P. L. Kuhns, A. P. Reyes, B. Liang, and C. T. Lin, Critical point and the nature of the pseudogap of single-layered copper-oxide $Bi_2Sr_{2-x}La_xCuO_{6+\delta}$ superconductors, *Phys. Rev. Lett.* **94**, 047006 (2005).

[181] J. L. Cardy and E. Rabinovici, Phase structure of $Z(p)$ models in presence of the theta parameter, *Nucl. Phys.* **B205**, 1–16 (1982).

[182] G. 't Hooft, Topology of the gauge condition and new confinement phases in non-Abelian gauge theories, *Nucl. Phys.* **190**, 455–478 (1981).

[183] M. A. Metlitski, C. L. Kane, and M. P. A. Fisher, Bosonic topological insulator in three dimensions and the statistical Witten effect, *Phys. Rev.* **B88**, 035131 (2013).

[184] R. Jackiw, Topics in planar physics, in *Physics, Geometry and Topology*, H. C. Lee (ed.), Plenum Press, New York (1989).

[185] C. Kittel, *Introduction to Solid State Physics*, Wiley, New York (1996).

[186] M. Sun, K. H. A. Villegas, V. M. Kovalev, and I. G. Savenko, Unconventional Bloch-Grüneisen scattering in hybrid Bose–Fermi systems, *Phys. Rev. Lett.* **123**, 095301 (2019); M. Sun, K. H. A. Villegas, V. M. Kovalev, and I. G. Savenko, Bogolon-mediated electron scattering in graphene and hybrid Bose Fermi systems, *Phys. Rev.* **B99**, 115408 (2019).

[187] V. J. Emery and S. A. Kivelson, Importance of phase fluctuations in superconductors with small superfluid density, *Nature* **374**, 434–437 (1995).

[188] P. Anderson, The resonating valence bond state in La_2CuO_4 and superconductivity, *Science* **235**, 1196 (1987).

[189] M. Harland, M. I. Katsnelson, and A. I Lichtenstein, Plaquette valence bond theory of high-temperature superconductivity, *Phys. Rev.* **B94**, 125133 (2016).

[190] G. Baskaran, Resonating valence bond theory of superconductivity: Beyond the cuprates, arXiv:1709.10070 (2017).

[191] J. R. Schrieffer, X. G. Wen, and S. C. Zhang, Dynamic spin fluctuations and the bag mechanism of high-T_c superconductivity, *Phys. Rev.* **B39**, 11663 (1989).

[192] A. V. Chubukov, A. Finkel'stein, R. Haslinger, and D. Morr, First-order superconducting transition near a ferromagnetic quantum critical point, *Phys. Rev. Lett.* **90**, 077002 (2003).

[193] A. V. Chubukov and J. Schmalian, Superconductivity due to massless boson exchange in the strong-coupling limit, *Phys. Rev.* **B72**, 174520 (2005).

[194] A. Bussmann-Holder and H. Keller, High-temperature superconductors: Underlying physics and applications, *Z. Naturforsch.* (2019).

[195] T. Shibauchi, T. Hanaguri, and Y. Matsuda, Exotic superconducting states in FeSe-based materials, *J. Phys. Soc. Jpn.* **89**, 102002 (2020).

[196] C. L. Kane and E. J. Mele, Size, shape and low-energy electronic structure of carbon nanotubes, *Phys. Rev. Lett.* **78**, 1932 (1997).

[197] F. Guinea, B. Horovitz, and P. L. Doussal, Gauge fields induced by ripples in graphene, *Phys. Rev.* **B77**, 041403 (2008).

[198] M. A. H. Vozmediano, M. I. Katsenlson, and F. Guinea, Gauge fields in graphene, *Phys. Rep.* **109**, 496 (2010).

[199] L.-C. Liu, Pseudo-magentic fields of strongly-curved graphene nanobubbles, *Int. J. Mod. Phys.* **B32**, 1850137 (2018).

[200] A. Pomar, M. V. Ramallo, J. Mosqueira, C. Torrón, and F. Vidal, Fluctuation-induced in-plane conductivity, magnetoconductivity, and diamagnetism of $Bi_2Sr_2CaCu_2O_8$ single crystals in weak magnetic fields, *Phys. Rev.* **B54**, 7470–7480 (1996).

[201] F. Zhao *et al.*, Sign-reversing Hall effect in atomically thin high-temperature $Bi_{2.1}Sr_{1.9}CaCu_{2.0}O_{8+\delta}$ superconductors, *Phys. Rev. Lett.* **122**, 247001 (2019).

[202] A. I. Baz, A. M. Perelomov, and Y. B. Zel'dovich, Scattering, reactions and decay in non-relativistic quantum mechanics, *Jerusalem: Israel Program for Scientific Translations*, 246 (1969).

[203] D. R. Nelson and V. M. Vinokur, Boson localization and correlated pinning of superconducting vortex arrays, *Phys. Rev.* **B48**, 13060–13097 (1993).

[204] E. Snider *et al.*, Room-temperature superconductivity in a carbonaceous sulfur hydride, *Nature* **586**, 373–377 (2020).

[205] L. B. Ioffe and A. I. Larkin, Properties of superconductors with a smeared transition temperature, *Sov. Phys. JETP* **54**, 378–384 (1981).

[206] E. Fradkin, *Field Theories of Condensed Matter Physics*, 2th edn, Cambridge University Press (2013).

[207] L. N. Bulaevskii, C. D. Batista, M. V. Mostovoy, and D. I. Khomskii, Electronic orbital currents and polarization in Mott insulators, *Phys. Rev.* **B78**, 024402 (2008).

[208] M. E. Simon and C. M. Varma, Detection and implications of a time-reversal breaking state in underdoped cuprates, *Phys. Rev.* **B89**, 247003 (2012).

[209] C. M. Varma, Pseudogap and Fermi-arcs in underdoped cuprates, *Phys. Rev.* **B99**, 224516 (2019).

[210] P. Bourges, D. Bounoua, and Y. Sidis, Loop currents in quantum matter, arxiv:2013.13295 (2021).

[211] A. Nahum, *Critical Phenomena in Loop Models*, Springer Verlag (2015).

[212] D. M. Basko, I. L. Aleiner, and B. L. Altshuler, Metal–insulator transition in weakly interacting many-electron system with localized single-particle states, *Ann. Phys.* **321**, 1126 (2006).

[213] D. A. Abanin, E. Altman, I. Bloch, and M. Serbyn, Many-body localization, thermalization and entanglement, *Rev. Mod. Phys.* **91**, 021001 (2019).

[214] V. Ros, M. Müller, and A. Scardicchio, Integrals of motion in the many-body localized phase, *Nucl. Phys.* **B891**, 420–465 (2015).

[215] M. Kormos, M. Collura, G. Takacs, and P. Calabrese, Real time confinement following a quantum quench to a non-integrable model, *Nat. Phys.* **13**, 246–249 (2016).

[216] R. C. Myers, M. Rozali, and B. Way, Holographic quenches in a confined phase, *J. Phys.* **A50**, 494002 (2017).

[217] M. Brenes, M. Dalmonte, M. Heyl, and A. Scardicchio, Many-body localization dynamics from gauge invariance, *Phys. Rev. Lett.* **120**, 030601 (2018).

[218] P. P. Mazza, G. Perfetto, A. Lerose, M. Collura, and A. Gambassi, Suppression of transport in non-disordered quantum spin chains due to confined excitations, *Phys. Rev.* **B99**, 180302 (2019).

[219] J. Park, Y. Kuno, and I. Ichinose, Glassy dynamics from quark confinement: Atomic quantum simulation of the gauge-Higgs model on a lattice, *Phys. Rev.* **A100**, 013629 (2019).

[220] A. J. A. James, R. M. Konik, and N. J. Robinson, Nonthermal tates arising from confinement in one and two dimensions, *Phys. Rev. Lett.* **122**, 130603 (2019).

[221] N. J. Robinson, A. J. James, and R. M. Konik, Signatures of rare states and thermalization in a theory with confinement, *Phys. Rev.* **B99**, 195108 (2019).

[222] A. Cortes Cubero and N. J. Robinson, Lack of thermalization in $(1+1)$-d QCD at large N_c, arXiv:1908.00270 (2019).

[223] T. Chanda, J. Zakrzewski, M. Lewenstein, and L. Tagliacozzo, Confinement and lack of thermalization after quenches in the bosonic Schwinger model, *Phys. Rev. Lett.* **124**, 1806012 (2020).

[224] A. Lerose, F. M. Surace, P. P. Mazza, G. Perfetto, M. Collura, and A. Gambassi, Quasilocalized dynamics from confinement of quantum excitations, *Phys. Rev.* **B102**, 041118(R) (2020).

[225] G. Giudici, F. M. Surace, J. E. Ebot, A. Scardicchio, and M. Dalmonte, Breakdown of ergodicity in disordered U(1) lattice gauge theories, *Phys. Rev. Research* **2**, 032034 (2020).

[226] Z. Yao, C. Liu, P. Zhang, and H. Zhai, Many-body localization from dynamical gauge fields, *Phys. Rev.* **B102**, 104302 (2020).

[227] B. Julia and G. Toulouse, The many-defect problem: Gauge-like variables for ordered media containing defects, *J. Phys. Lett.* **40**, 395–398 (1979).

[228] C. Nash and S. Sen, *Topology and Geometry for Physicists*, Academic Press, London (1983).

[229] R. Rajaraman, *Solitons and Instantons*, North-Holland, Amsterdam (1987).

[230] E. C. G. Stückelberg, Radiation theory for photons of arbitrarily small mass, *Helv. Phys. Acta* **30**, 209 (1957).

[231] S. Pu, G. J. Sreejith, and J. K. Jain, Anderson localization in fractional quantum Hall effect, arXiv:2109.00362 (2021).

[232] D. M. Basko, I. L. Aleiner, and B. L. Altshuler, Possible experimental manifestations of the many-body localization, *Phys. Rev.* **B76**, 052203 (2007).

[233] B. L. Altshuler, V. E. Kravtsov, I. V. Lerner, and I. L. Aleiner, Jumps in current-voltage characteristics in disordered films, *Phys. Rev. Lett.* **102**, 176803 (2009)

[234] I. Tamir, T. Levinson, F. Gorniaczyk, A. Doron, J. Lieb, and D. Shahar, Excessive noise as test for many-body localization, *Phys. Rev.* **B99**, 035135 (2019).